Genome analysis

a practical approach

TITLES PUBLISHED IN
THE
PRACTICAL APPROACH
SERIES

Series editors:
Dr D Rickwood
Department of Biology, University of Essex
Wivenhoe Park, Colchester, Essex CO4 3SQ, UK
Dr B D Hames
Department of Biochemistry, University of Leeds
Leeds LS2 9JT, UK

Affinity chromatography

Animal cell culture

Antibodies

Biochemical toxicology

Biological membranes

Carbohydrate analysis

Centrifugation (2nd Edition)

DNA cloning

Drosophila

Electron microscopy
in molecular biology

Gel electrophoresis of nucleic acids

Gel electrophoresis of proteins

Genome analysis

HPLC of small molecules

Human cytogenetics

Human genetic diseases

Immobilised cells and enzymes

Iodinated density gradient media

Lymphocytes

Lymphokines and interferons

Mammalian development

Microcomputers in biology

Mitochondria

Mutagenicity testing

Neurochemistry

Nucleic acid and
protein sequence analysis

Nucleic acid hybridisation

Oligonucleotide synthesis

Photosynthesis:
energy transduction

Plant cell culture

Plasmids

Prostaglandins
and related substances

Spectrophotometry
and spectrofluorimetry

Steroid hormones

Teratocarcinomas
and embryonic stem cells

Transcription and translation

Virology

Yeast

Genome analysis

a practical approach

Edited by
K E Davies

Molecular Genetics Group, Nuffield Department of Clinical
Medicine, John Radcliffe Hospital, Oxford OX3 9DU, UK

OXFORD · WASHINGTON DC

IRL Press
Eynsham
Oxford
England

British Library Cataloguing in Publication Data
Genome analysis.
  1. Organisms. Genes. Mapping
  I. Davies, K.E. (Kay E.)   II. Series
  575.1′2
Library of Congress Cataloging-in-Publication Data
Genome analysis: a practical approach / edited by K.E. Davies.
p. cm. — (Practical approach series)
Includes bibliographies and index.
1. Chromosome mapping — Technique. I. Davies, Kay. II. Series.
[DNLM: 1. Chromosome Mapping. QH 445.2 G335]
QH445.2.G465 1988
575. 1′2 — dc19   88-23023
ISBN 1 85221 109 1 $54.00 (U.S.).
ISBN 1 85221 110 5 (Pbk.) : $36.00 (U.S.)

Typeset by Infotype and printed by Information Printing Ltd, Oxford, England.

Preface

Methodologies used to analyse the genomes of prokaryotes and eukaryotes are advancing very rapidly as it has been realized that there is a resolution gap between what can be viewed under the microscope and what molecular techniques can resolve. Obviously, a valuable goal is to obtain the complete sequence of the genome of interest although even this will not supply all the answers to the control of gene expression. This volume aims to present current successful strategies which are being applied to various organisms to construct physical maps in order to identify and analyse the arrangement and function of genes. These techniques are being used to move from linked markers to candidate genes in several human monogenic disorders. Such advances in the isolation of important disease loci have been paralleled by impressive developments in the sensitivity of tests used to detect and analyse human mutations. The style of presentation of methods in this book should enable any research or diagnostic laboratory to apply them to their particular systems. Finally, two chapters are also included describing new probes and approaches for the localization of human disorders whose analysis has so far eluded the molecular geneticist.

K.E.Davies

Contributors

G.S.Banting
Human Molecular Genetics Laboratory, Imperial Cancer Research Fund, Lincoln's Inn Fields, London WC2A 3PX, UK

C.R.Cantor
Department of Genetics and Development, College of Physicians and Surgeons, Columbia University, New York City, NY 10032, USA

F.S.Collins
Division of Medical Genetics, The University of Michigan and The Howard Hughes Medical Institute, Ann Arbor, MI 48109, USA

A.Coulson
Medical Research Council, Laboratory of Molecular Biology, Hills Road, Cambridge CB2 2QH, UK

D.R.Cox
Departments of Pediatrics and Biochemistry and Biophysics, University of California at San Francisco, 513 Parnassus Street, San Francisco, CA 94143, USA

H.A.Erlich
Department of Human Genetics, Cetus Corporation, 1400 Fifty-third Street, Emeryville, CA 94608, USA

P.N.Goodfellow
Human Molecular Genetics Laboratory, Imperial Cancer Research Fund, Lincoln's Inn Fields, London WC2A 3PX, UK

U.B.Gyllensten
Department of Human Genetics, Cetus Corporation, 1400 Fifty-third Street, Emeryville, CA 94608, USA

S.R.Klco
Department of Genetics and Development, College of Physicians and Surgeons, Columbia University, New York City, NY 10032, USA

E.S.Lander
Whitehead Institute of Biomedical Research, 9 Cambridge Center, Cambridge, MA 02142, USA

R.M.Myers
Departments of Physiology and Biochemistry and Biophysics, University of California at San Francisco, 513 Parnassus Street, San Francisco, CA 94143, USA

C.A.Pritchard
Human Molecular Genetics Laboratory, Imperial Cancer Research Fund, Lincoln's Inn Fields, London WC2A 3PX, UK

R.K.Saiki
Department of Human Genetics, Cetus Corporation, 1400 Fifty-third Street, Emeryville, CA 94608, USA

V.C.Sheffield
Departments of Pediatrics and Biochemistry and Biophysics, University of California at San Francisco, 513 Parnassus Street, San Francisco, CA 94143, USA

C.L.Smith
Departments of Microbiology and Psychiatry, College of Physicians and Surgeons, Columbia University, New York City, NY 10032, USA

J.Sulston
Medical Research Council, Laboratory of Molecular Biology, Hills Road, Cambridge CB2 2QH, UK

R.A.Wells
Nuffield Department of Clinical Medicine, John Radcliffe Hospital, Headington, Oxford OX3 9DU, UK

Contents

5. DETECTION OF SINGLE BASE CHANGES IN DNA: RIBONUCLEASE CLEAVAGE AND DENATURING GRADIENT GEL ELECTROPHORESIS   95

R.M.Myers, V.C.Sheffield and D.R.Cox

6. THE POLYMERASE CHAIN REACTION   141

R.K.Saiki, U.B.Gyllensten and H.A.Erlich

Abbreviations

AMV	avian myeloblastosis virus
BSA	bovine serum albumin
CMGT	chromosome-mediated gene transfer
DGGE	denaturing gradient gel electrophoresis
DMEM	Dulbecco's modified Eagle's medium
DMGT	DNA-mediated gene transfer
DMSO	dimethyl sulphoxide
dNTP	deoxynucleotide triphosphate
DTT	dithiothreitol
EC	embryonal carcinoma
EDTA	ethylenediamine tetraacetic acid
FACS	fluorescence activated cell sorter
FCS	fetal calf serum
FTL	freeze−thaw lysate
HAT	hypoxanthine, aminopterin, thymidine
HBS	Hepes-buffered saline
Hepes	N-hydroxyethylpiperazine-N'-2-ethanesulphonic acid
HPRT	hypoxanthine phosphoribosyl transferase
HVR	hypervariable region
IFGT	irradiation and fusion gene transfer
IPTG	isopropyl-1-thio-β-D-galactoside
MMGT	microcell-mediated gene transfer
NF	neurofibromatosis
NTP	nucleotide triphosphate
PBS	phosphate-buffered saline
PCR	polymerase chain reaction
PEG	polyethylene glycol
PFG	pulsed-field gel
PHA	phytohaemagglutinin
Pipes	piperazine-N,N'-bis-2-ethanesulphonic acid
PMSF	phenylmethylsulphonyl fluoride
RFLP	restriction fragment length polymorphism
SDS	sodium dodecyl sulphate
SE	sonicated extract
SSC	standard saline citrate
TAE	Tris−acetate−EDTA buffer
TBE	Tris−borate−EDTA buffer
TE	Tris−EDTA buffer
TEMED	(N,N,N',N'-)tetramethylethylenediamine
TK	thymidine kinase
YAC	yeast artificial chromosome

CHAPTER 1

Techniques for mammalian genome transfer

P.N.GOODFELLOW, C.A.PRITCHARD and G.S.BANTING

1. INTRODUCTION

All genome transfer experiments have two separate components: a method for trans-
ferring genetic material from donor to recipient cell and a selection method for isolating
recipient cells that have received donor material. Most transfer methods result in between
1 in 10^3 and 1 in 10^7 ($10^{-3} - 10^{-7}$) recipient cells receiving the desired material,
consequently very powerful selection techniques are required and frequently the selection
is the limiting component in a genome transfer experiment.

The first genome transfers were performed by whole cell fusion (1). This technology
was applied to the study of differentiation and the tumour phenotype, however, the
most successful applications have been in human gene mapping (2) and the production
of monoclonal antibodies (3). If a rodent cell is fused with a human cell, the resultant
interspecific hybrid spontaneously loses human chromosomes (4). The chromosome
loss is essentially random and this allows the construction of hybrid cell lines containing different human chromosome contributions. The correlation between the presence
in a hybrid of a human chromosome and the presence of a human genetic marker is
the basis of human gene mapping by somatic cell genetics. Of the 1300 human genes
assigned to specific chromosomes over one third have been mapped by somatic cell
genetic methods (5). In contrast to the chromosomal instability of interspecific hybrids,
the loss of chromosomes from intraspecific hybrids is modest (6). The fusion of a mouse
myeloma with mouse spleen cells produces stable hybrid cell lines with the immortalized growth characteristics of the myeloma parent and the antibody secreting properties of the spleen cells. The *in vitro* production of monoclonal antibodies has
revolutionized antibody-based research as well as diagnostic, therapeutic and industrial
methods.

The chromosomal instability of interspecific hybrids and the chromosomal complexity
of tetraploid intraspecific hybrids has limited the genetic analysis of complex phenotypes
by somatic cell genetics. Microcell-mediated gene transfer (MMGT) was developed
to allow the transfer of single chromosomes between somatic cells (7). This method
has proved to be technically difficult, nevertheless MMGT has been used to construct
hybrids for gene mapping and analysis of differentiation and cancer. The latter
experiments have provided conclusive evidence for the existence of specific *trans*-acting
regulators of the differentiated phenotype (8) and have confirmed the recessive properties
of the Wilm's tumour gene (9).

Although whole cell fusion and MMGT have been exploited for chromosomal localization of genes, subchromosomal localization by these techniques has been restricted

by the availability of translocation and deletion chromosomes. Two experimental solutions to this problem have been attempted: chromosome-mediated gene transfer (CMGT) (10) and irradiation and fusion gene transfer (IFGT) (11). In CMGT purified mitotic chromosomes are added to recipient cells in the presence of calcium phosphate. The recipient cells incorporate fragments of donor chromosomes into their genomes. Application of suitable selection systems allows the isolation of hybrids containing donor fragments of interest. Unfortunately, the chromosome fragments are frequently rearranged and, in some cases, there is clear evidence for non-random retention of sequences from the centromeric region (12). These observations preclude the use of CMGT as a mapping technique, nevertheless CMGT has proved to be useful for the enrichment of specific chromosomal regions as part of cloning strategies employing reverse genetics. Goss and Harris (11) were the first to demonstrate that fragments of the human X chromosome, present in γ-irradiated human cells, could be rescued by fusion of the donor cells to rodent recipient cells. In a recent re-investigation of IFGT, we have found that the fragments generated are rearranged and show evidence for the same centromeric selection observed with CMGT (unpublished observations).

Genomic DNA can be transferred by adding purified DNA to recipient cells in DNA-mediated gene transfer (DMGT). This technique is also an important method for testing gene function in mammalian cells. Three main methods for DMGT have been developed: calcium phosphate precipitation (13), electroporation (14) retroviral vector-mediated transfer (15) and microinjection (16). It is generally assumed that for each of these techniques the size of the DNA fragment transferred intact is limited. This assumption has not been rigorously tested for all the methods but it is not expected that fragments greater than 100 kb will be routinely transferable. In addition to its use in testing gene function, DMGT has been used for the random incorporation of selectable genes into the genome (17) and the cloning of selectable genes by expression (18).

In this chapter we describe methods for the transfer of genomic DNA by whole cell fusion, MMGT, CMGT and DMGT. General reviews, which explore the biology and applications of these techniques can be found elsewhere (19−24).

2. GENERAL CONSIDERATIONS

2.1 **Good husbandry**

All cells cultured *in vitro* should be maintained and passaged under conditions which maximize cell viability. This self-evident edict is especially true of cells to be used in experimental manipulations. Most genome transfer experiments involve exposure to potentially toxic chemicals [e.g. polyethylene glycol (PEG)] and cells in poor condition are more likely to be killed.

Transfer techniques that require isolation of clones will be limited by the plating efficiency of the recipient cells. Consequently, all cell lines to be used as recipients should be checked for plating efficiency under the conditions of the experiment. Cell lines with plating efficiencies of less than 10% should not be used except *in extremis*.

All of the manipulations described in this chapter should be carried out using *sterile* techniques and reagents. All solutions should be made with tissue culture grade reagents and double-distilled water. It is assumed that the reader will be familiar with standard tissue culture and cell cloning methods.

2.2 **Selection**

It is outside the scope of this chapter to present details of selection methods; however, it should be stressed that without an appropriate selection method it is pointless to attempt a genome transfer experiment. Selection methods are designed to confer a growth advantage on the desired cell. The commonly used methods can be divided into four different categories.

2.2.1 *Biochemical selection based on endogenous genes*

The most widespread example of this approach is the hypoxanthine, aminopterin, thymidine (HAT) selection methods of Szybalski *et al.* (25). In the presence of aminopterin (or the closely related drug methotrexate) *de novo* synthesis of DNA precursors is inhibited. Cells that lack the enzyme thymidine kinase (TK) cannot utilize exogenous thymidine and die in the presence of aminopterin. Similarly, cells that lack hypoxanthine phosphoribosyl transferase (HPRT) cannot utilize hypoxanthine and also die in the presence of aminopterin. The somatic cell hybrid produced by fusing a TK^-, $HPRT^+$ with a TK^+ $HPRT^-$ will be both $HPRT^+$ and TK^+. This hybrid cell will grow in the presence of aminopterin if hypoxanthine and thymidine are supplied (HAT medium). A common variant of the HAT theme is 'half selection' where one cellular partner does not grow in culture and the other is HPRT or TK deficient (26). A large number of complementation systems have been developed for somatic cell genetics; particularly numerous are those based on the auxotrophy/prototrophy of hamster cell lines (27). The problem with this approach is that it requires the construction of suitable mutant recipient cells.

In view of the general importance of the HAT selection method we have presented the specific details of making HAT medium in *Table 1*.

Table 1. Preparation of HAT medium.

HAT medium is made by adding 1 ml of solution 1 and 1 ml of solution 2 to 98 ml of growth medium.

Solution 1: methotrexate (also known as amethopterin)

1.	Add 0.045 g to 10 ml of double-distilled water.
2.	Add 1.0 M sodium hydroxide until the methotrexate dissolves.
3.	Add 10 ml of double-distilled water.
4.	Adjust the pH to between 7.5 and 7.8 with 1 M HCl.
5.	Make up to 100 ml and filter sterilize.
6.	Store the stocks frozen.

Solution 2: hypoxanthine and thymidine

1.	Add 0.14 g of hypoxanthine to 30 ml of double-distilled water.
2.	Add 1 M sodium hydroxide until the hypoxanthine dissolves.
3.	Adjust the pH to 10.0 with 1 M HCl.
4.	Add 0.039 g of thymidine to 35 ml of double-distilled water.
5.	Combine the hypoxanthine and thymidine solutions and adjust to 100 ml using double-distilled water.
6.	Filter sterilize and store frozen ($-20°C$).

2.2.2 *Biochemical selection of exogenous genes*

This approach avoids the need to isolate specific mutants by exploiting bacterial genes that confer selective advantage when expressed in mammalian cells (28). Plasmid and retroviral expression vectors have been created in which the bacterial genes have been given mammalian promoters, splice sites and polyadenylation signals. Introduction of the bacterial genes into mammalian cells by transfection or infection results in random integration into the recipient genome. Examples of genes that provide a dominant selective advantage to mammalian cells include the *Escherichia coli gpt* gene which allows cells to utilize xanthine as a precursor for purine synthesis (28) and the *neo* gene which confers resistance to the mammalian-toxic antibiotic G418 (29). The major disadvantage with this approach is the random integration; however, recent advances in targeting by homologous recombination may allow site-directed integration (30).

2.2.3 *Cell surface antigens*

Antibodies can be used as selective agents for isolating cells with defined cell surface antigen phenotypes (31). A number of different methods are available including the use of the fluorescence-activated cell sorter (FACS), rosetting with antibody-coupled red cells and panning on antibody-coated surface. Selection of antigen-positive cells is generally limited in efficiency and is more correctly described as enrichment. Nevertheless, antibody-based selection has been used with all of the genome transfer methods described here, with the exception of MMGT.

2.2.4 *Activated oncogenes*

Mutated proto-oncogenes, especially members of the *ras* family, can confer growth advantages on mammalian cells and this can be exploited for selection in somatic cell genetics (22).

3. WHOLE CELL FUSION

3.1 **Introduction**

Cells mixed together will spontaneously fuse, but only with a very low efficiency (1). Fusion rates can be markedly increased by fusogenic agents. Initial experiments used inactivated Sendai virus as the fusogen (32), however, biological variability in different batches of inactivated virus and the cumbersome nature of the virus-promoted fusion protocols has resulted in the widespread use of the chemical fusogen PEG (33). The method presented below can be used for fusing cells of the same or different species. Intraspecific hybrids should be produced with a frequency of between 10^{-3} and 10^{-5}. Interspecific hybrids should be produced with a frequency of between 10^{-5} and 10^{-7}. Cells with similar phenotypes and growth characteristics tend to produce hybrids with a higher frequency.

3.2 **Basic method for producing whole cell hybrids by PEG-induced fusion**

(i)　　Prepare solutions listed in *Table 2*.

(ii)　　Harvest 5×10^6 of each parental cell, transfer to a sterile centrifuge tube with a conical base and spin to a common pellet by centrifugation (1500 *g* for 5 min

4

Table 2. Stock solutions required for whole cell fusion.

Growth medium	Routinely we use Dulbecco's modified Eagle's medium (DMEM) supplemented with 10% FCS for cell culture, however, the particular growth medium used does not directly affect the fusion protocol.
Growth medium without FCS (or other proteins)	
Selective medium	Growth medium suitable for selecting hybrid cells (e.g. HAT medium for isolating hybrids between TK$^-$ and HPRT$^-$).
50% v/v PEG made up in serum-free medium	5.5 g of PEG 4000 (Baker) and 5 ml of serum-free growth medium are mixed and autoclaved. The final pH should be adjusted to 8.2 and the solution should be pre-warmed to 37°C immediately prior to use.

at room temperature). Wash the pellet once by resuspension and centrifugation in serum-free medium.

(iii) Remove the supernatant and loosen the pellet by tapping the centrifuge tube.

(iv) Gently add 1 ml of pre-warmed 50% v/v PEG (37°C) whilst carefully disturbing the pellet with the tip of the pipette used to add the PEG (see below for a discussion of PEG grade and supplier).

(v) Incubate for 90 sec at 37°C.

(vi) Add 1 ml of serum-free medium drop by drop over a period of 1 min. Carefully mix the contents of the tube by gentle stirring with the pipette used to add the serum-free medium.

(vii) Add 5 ml of serum-free medium over a period of 1−2 min. Mix the contents of the tube by gentle stirring with the pipette.

(viii) Add 10 ml of serum-supplemented medium over a period of 1−2 min. Mix the contents of the tube by gentle stirring with the pipette.

(ix) Centrifuge the cells (1500 g for 5 min at room temperature), remove the supernatant and gently resuspend the cells in growth medium by gentle tapping of the tube. Do not pipette the cells up and down to resuspend the pellet. Plate the cells in 10 × 9.0 cm diameter tissue culture plates.

(x) After 24 h change the growth medium to selective growth medium. Change the medium every 3−4 days.

(xi) Colonies should be visible in 14−21 days for intraspecific hybrids and 21−28 days for interspecific hybrids. Occasionally hybrids may take longer to appear and it is worth keeping plates for up to 45 days.

3.3 Trouble-shooting and variant techniques

Some combinations of cells are difficult to fuse, if it proves impossible to produce hybrids, it is worth trying the following:

(i) Vary the ratios of the parental cells over the range 1:10−10:1.

(ii) Vary the PEG solution.

　　(a) The batch and/or supplier. In our experience a batch that works well for one cell type tends to work for other cells. A batch which is known to induce fusion with a high frequency should be guarded. It has been claimed that

5

Table 3. A PEG fusion method for attached cells.

1.	Prepare the solutions listed in *Table 2*.
2.	Plate 2.5×10^5 cells of each parental line in a 9.0 cm diameter tissue culture dish (use as many dishes as required).
3.	Culture until the cells are confluent (1−2 days).
4.	Remove the medium and wash twice in serum-free medium.
5.	Add 1.5 ml of PEG solution and tilt the plate until the PEG solution has covered the whole surface.
6.	Leave for 90 sec at 37°C.
7.	Remove the PEG solution and gently wash twice with serum-free medium.
8.	Wash with medium containing serum.
9.	Proceed as for steps x and xi in the basic protocol.

poor batches of PEG can be improved by using solutions which do not contain calcium (34).

(b)   The molecular weight grade. As a rule the higher the molecular weight the greater the fusogenic activity. However, the higher molecular weight grades are more viscous and more difficult to wash away from the cells. As PEG is toxic, delay in removal may kill cells and reduce the number of hybrids obtained. We have successfully produced hybrids with PEG in the range 1000−6000 average molecular weight.

(c)   The concentration of PEG. Lower concentrations of PEG are less toxic and less fusogenic.

(iii)   Instead of incubating the cells in PEG for 90 sec at 37°C (step v in the basic protocol), centrifuge the cells in PEG for 2 min at 1500 *g* at room temperature; then proceed with steps vi−xi.

If these variations fail to yield hybrids it may be worth attempting alternative fusion protocols. One method, suitable for attached cells, is given in *Table 3*. This method can also be modified for use with an attached cell line to be fused with a suspension cell line. The attached cell line is cultured to subconfluence and treated with PEG (steps 1−4, *Table 3*), the suspension cells (5×10^5) are added in a small volume (<1.0 ml) of serum-free medium and then the plates are centrifuged for 2 min at 1000 *g* using a plate holding rotor. Subsequent treatment follows steps 6−8 in *Table 3*. An alternative to centrifugation is to use lectins to agglutinate the attached and non-attached cells.

If, after exhausting these possibilities, success is still elusive, a change of fusogen might help. Two obvious alternatives are Sendai virus and electroporation. We have no experience with using the latter technique, but we are willing to provide, upon request, laboratory protocols for use with Sendai virus.

4. MICROCELL-MEDIATED GENE TRANSFER

4.1 Introduction

A summary of the MMGT strategy is presented in *Figure 1*. Donor cells are blocked in mitosis by treatment with colcemid. After extended treatment the donor cells re-form the nuclear membrane around single or small groups of chromosomes. This results in the formation of cells containing multiple micronuclei. Enucleation of the cells by treatment with cytochalasin B and centrifugation produces microcells (micronuclei

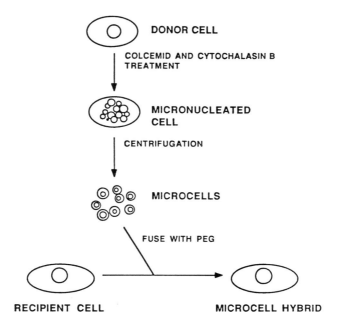

Figure 1. Microcell-mediated gene transfer.

encapsulated in plasma membrane). The microcells are delicate, but can be fused to recipient cells using a slightly modified PEG fusion protocol. Initial experiments suggested that only rodent cells could be induced to produce micronuclei, however, with variation in colcemid dose and length of exposure, most cells will produce micronuclei.

The MMGT procedure is finicky, time-consuming and technically difficult. In our experiments microcell hybrids are produced at a frequency of only $10^{-6} - 10^{-7}$ in interspecific fusions.

4.2 Method for MMGT

Day 1

(i)   Prepare stock solutions (*Tables 2, 4* and *5*).

(ii)   Preparation of donor cells for microcelling. Plate out 2×10^7 cells at 50% confluence and grow overnight in growth medium containing 0.05 μg/ml colcemid. This will induce microcell formation (7). For different cell lines it may be necessary to titrate both the concentration of colcemid (try $0.02 - 0.10$ μg/ml) and length of exposure (try $12 - 48$ h). Micronucleation can be easily scored by microscopic examination using phase contrast optics. In a good experiment about 50% of the cells will be micronucleated. This frequency increases after the cytochalasin B treatment.

Table 4. Stock solutions required for MMGT.

1.	50% v/v Ficoll in double-distilled water. This solution takes up to 12 h to dissolve. Sterilize by autoclaving.
2.	1 mg/ml cytochalasin B (Sigma) in dimethyl sulphoxide (DMSO).
3.	10 × concentrated Hank's buffered salts solution (Gibco).
4.	10 × concentrated sodium bicarbonate solution: 3.5 g/l in double-distilled water.
5.	Double-distilled water.
6.	The solutions described in *Table 2*.
7.	10 mg/ml colcemid (Difco) in distilled water.
8.	10 mg/ml phytohaemagglutinin A-P, (PHA-P; Difco) in distilled water.

Table 5. Preparation of Ficoll gradient solutions for MMGT.

Final Ficoll concentration (%)	50% Ficoll (ml)	Hank's (×10) (ml)	Bicarbonate (×10) (ml)	Water (ml)	Cytochalasin B (μl)
25	5.0	1.0	1.0	3.0	50
17	3.4	1.0	1.0	4.6	50
16	3.2	1.0	1.0	4.8	50
15	3.0	1.0	1.0	5.0	50
12.5	2.5	1.0	1.0	5.5	50
10	10.0	5.0	5.0	20.0	250
0	0	3.0	3.0	24.0	150

The gradient solutions are prepared and stored at 4°C using the solutions described in *Table 4*.

Day 2

(iii)   16 h later replace the medium with medium containing cytochalasin B at 2 μg/ml; incubate overnight.

(iv)   Transfer the gradient stock solutions (*Table 5*) to a 5% CO_2:95% air, 37°C incubator with the tops of the bottles slightly loosened. Incubate overnight to equilibrate.

(v)   Pre-warm a Beckmann SW41 rotor to 37°C by incubation overnight at 37°C.

Day 3

(vi)   Preparation of Ficoll step gradients. Thoroughly rinse the required number of Beckmann SW41 centrifuge tubes with 100% ethanol, dry by inversion in a laminar flow hood. Carefully pour the gradients using equilibrated stock solutions (*Figure 2*).

(vii)   Harvest colcemid- and cytochalasin-treated cells and pellet the cells by centrifugation (1500 *g* for 5 min at room temperature). Resuspend in 3 ml of 10% Ficoll solution (*Table 5*) and gently layer onto the prepared gradient. Fill the tubes with 0% Ficoll solution.

(viii)   Load the tubes containing the gradients into the pre-warmed (37°C) Beckmann SW41 rotor and place the rotor into a pre-warmed (37°C) ultracentrifuge. (The SW41 is a swing-out rotor.)

(ix)   Centrifuge for 1 h at 25 000 r.p.m. (80 000 *g*) at 37°C using minimum rates of acceleration and deceleration to avoid disruption of the gradients.

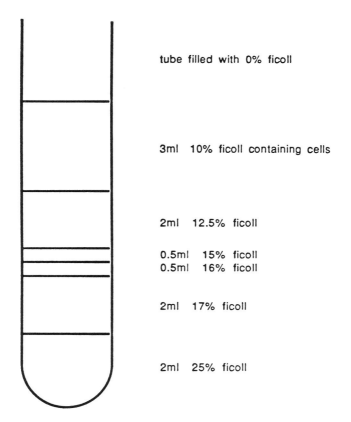

tube filled with 0% ficoll

3ml   10% ficoll containing cells

2ml   12.5% ficoll

0.5ml   15% ficoll
0.5ml   16% ficoll

2ml   17% ficoll

2ml   25% ficoll

Figure 2. Ficoll gradients for MMGT.

(x)   Remove the tubes from the centrifuge. Microcells should appear as fuzzy bands at 15:16% and 16:17% interfaces. Carefully remove these bands using a sterile Pasteur pipette introduced from the top of the gradient; transfer to a fresh, sterile SW41 tube and fill the tube with growth medium. Contamination with cytoplasts, karyoplasts and whole cells can be monitored by phase contrast microscopy.

(xi)   Centrifuge at 20 000 r.p.m. (50 000 g) at room temperature in a SW41 rotor for 10 min with maximum acceleration and deceleration. This pellets the microcells and rinses them free of Ficoll.

(xii)   Three different procedures for microcell purification are given at the end of this method. If a suitable selection system is available it may be possible to avoid purification of microcells. In this case they can be fused directly to recipient cells.

(xiii) Fusion of microcells to whole cells. Harvest the recipient cells and wash them in serum-free medium. Add about 10^7 recipient cells to the microcell pellet in 2 ml of serum-free growth medium supplemented with 100 μg/ml phytohaemagglutinin. Transfer to a plastic Universal bottle with a conical base and incubate for 10 min at 37°C to clump.

(xiv)   Pellet by centrifugation (1500 g for 5 min at room temperature).

(xv)   Fuse the microcells to whole cells using PEG as described for whole cell fusions.

4.3 **Purification of microcells**

4.3.1 *Method A* (7)

(i)   Collect and pool bands containing microcells from the Ficoll gradient.
(ii)   Resuspend, and make up to 22 ml in 0.5% bovine serum albumin (BSA) in phosphate-buffered saline (PBS) (pH 7.2).
(iii) Pellet by centrifugation at 50 000 *g* for 10 min at room temperature.
(iv)   Resuspend the pellet in 2 ml of 0.5% BSA in PBS (pH 7.2) and put on a 50 ml 1−3% linear BSA gradient (pH 7.3) in PBS (the gradient is poured from 25 ml of 1% BSA and 25 ml of 3% BSA using a gradient maker); leave for 3.5 h at room temperature. Whole cells will sink to the bottom of the gradient, microcells will remain near the top.
(v)   Collect the top 20 ml (excluding sample zone) of the gradient and pellet by centrifugation (1500 *g* for 5 min at room temperature).
(vi)   Fuse the pelleted microcells to recipient cells using the PEG protocol.

4.3.2 *Method B* (35)

(i)   Collect and pool the bands containing microcells from the Ficoll gradient.
(ii)   Resuspend and make up to 22 ml in growth medium.
(iii) Pellet by centrifugation at 50 000 *g* for 10 min at room temperature.
(iv)   Resuspend the pellet in 20 ml of growth medium and plate out in 2 × 9.0 cm diameter tissue culture dishes. Leave for 90 min at 37°C.
(v)   Remove the medium from the dishes and re-plate in fresh dishes. Leave for 90 min at 37°C.
(vi)   Remove the medium and spin the microcells to a common pellet with the cells to which they are to be fused by centrifugation (1500 *g* for 5 min at room temperature).
(vii) Fuse the microcells to cells using the PEG protocol.

4.3.3 *Method C* (36)

(i)   Collect and pool the bands containing microcells from the Ficoll gradient.
(ii)   Resuspend and make up to 20 ml in growth medium.
(iii) Pellet by centrifugation at 50 000 *g* for 10 min at room temperature. Resuspend in 10 ml of growth medium and pass through a 0.8 μm polycarbonate filter (Nucleopore). The filter is mounted in a swinnex adaptor and sterilized by autoclaving. It may be necessary to use several filters. The suspended microcells should not be forced through the filter; use only gentle pressure.
(iv)   Pass the filtrate through a 0.3 μm polycarbonate filter (Nucleopore).
(v)   Pellet the microcells by centrifuging the filtrate (1500 *g* for 5 min at room temperature).
(vi)   Fuse the microcells to recipient cells using the PEG protocol.

4.4 Trouble-shooting and variant protocols

Prolonged colcemid treatment followed by cytochalasin B treatment is the simplest method for inducing micronucleation (20), however it may be necessary to titrate both the concentration and time of treatment with both reagents. If this method proves unsatisfactory, two alternative strategies can be contemplated. It may be possible to produce an intermediary donor cell: in the past we have constructed human—rodent somatic cell hybrids and used these as donors of human chromosomes (37). The hybrid cells form micronuclei under the same conditions as the rodent parent. The second strategy is to use an alternative method of inducing micronuclei, for example cells can be micronucleated by treatment with nitrous oxide (38).

The enucleation procedure presented above is based on the method of Wigler and Weinstein (39). We have used this method because it can be used with both attached and unattached cells. In addition, this method can be scaled up for use with large numbers of cells. Most of our colleagues have eschewed this approach and have chosen instead to follow the protocols originally developed by Prescott *et al.* (40). Cells are cultured attached to coverslips or plastic bullets. The bullets are placed in centrifuge tubes and centrifuged in medium containing cytochalasin B. In some protocols whole tissue culture flasks are centrifuged at 20 000 *g*. It is claimed that these methods may produce preparations with less whole cell contamination.

The low frequency of the MMGT procedure means that whole cell contamination can cause problems even if a selection is available against whole donor cells. In some experiments we have recovered whole cell hybrids and even revertant donor whole cells which have escaped selection. The microcell purification procedures can reduce this problem but at the cost of reduced yield of microcells and consequent reduction in yield of microcell hybrids. One way of avoiding these problems is to use a recipient cell with a different morphological phenotype to the donor cells. In several experiments we have used mouse embryonal carcinoma (EC) cells as recipient cells. The morphology of these cells is recessive to L cells in whole cell fusions. In experiments in which human—rodent hybrids with an L cell morphology were used as human chromosome donors and EC cells as recipients, whole cell hybrids could be distinguished easily from microcell hybrids.

A final note of caution: 30% of our MMGT clones have contained rearranged chromosomes and 30% have contained more than one donor chromosome.

5. CHROMOSOME-MEDIATED GENE TRANSFER

5.1 Introduction

CMGT can be used to transfer fragments of chromosomes from the nucleus of one cell type to the nucleus of another cell type. Although, in theory, any cell type can be used as a chromosome donor or acceptor, in practice, the technique depends upon the availability of recipient cell lines which have the property of being highly transfectable. High transfection frequencies can be obtained using immortalized mouse cell lines as recipients. Lower transfection frequencies are usually found with hamster cell lines and immortalized human cell lines, although recent results have suggested that the human EJ bladder carcinoma-derived cell line can be transfected with chromosomes at a frequency comparable to mouse L cells (unpublished observations). In contrast to recipient cells, many different types of cells have been utilized as chromosome donors and we have obtained good chromosome yields from cell lines grown both in suspension and as attached monolayers. Cells that are easy to grow and handle in large quantities are usually preferred as donors. The following protocols describe general procedures for isolating donor chromosomes and for establishing fragments of these chromosomes in recipient cells using CMGT.

5.2 **Chromosome isolation**

In the following protocol it is important to use plastic pipettes and tubes, to prevent chromosome loss and to perform all manipulations at 4°C, so that chromosome disintegration is kept to a minimum. Cells are blocked in mitosis, mitotic chromosomes are released by hypotonic shock and shearing. Chromosomes are purified by differential centrifugation.

(i)   Prepare stock solutions (*Table 6*).
(ii)   Grow donor cells to confluence. This serves to partially synchronize the culture.
(iii)   Passage the cells 16−24 h before transfection. Plate the cells at 30% confluence in medium containing 20% fetal calf serum (FCS) and 0.1 μg/ml colcemid.
(iv)   The next day, harvest the cells and wash them twice in medium without serum (or other proteins).
(v)   Resuspend the cells in hypotonic buffer using 10 ml per 10^8 cells. Incubate at 4°C for 20−30 min. Pass the cell suspension through a 21-gauge needle 10 times, maintaining the suspension on ice throughout.
(vi)   Centrifuge the broken cells at 700 g for 7 min at 4°C to remove contaminating cells and nuclei.

Table 6. Stock solutions required for CMGT.

Growth and selective media	NB. The transfection must be performed in a low phosphate medium such as DMEM. For cells growing in high phosphate medium change to low phosphate medium immediately prior to transfection. The high phosphate medium can be used after the glycerol shock.
Hypotonic buffer (pH 7.1)	10 mM Hepes
	3 mM calcium chloride
Transfection buffer (pH 7.1; the pH is critical)	25 mM Hepes
	134 mM sodium chloride
	5 mM potassium chloride
	0.7 mM sodium phosphate (monobasic)
	5 mM glucose
1.25 M calcium chloride	
Wash buffer (pH 7.1)	25 mM Hepes
	134 mM sodium chloride
	5 mM potassium chloride
	0.7 mM sodium phosphate (monobasic)

(vii)   Collect the supernatant and pellet the chromosomes by centrifugation at 2500 *g* for 25 min at 4°C.
(viii) Resuspend the pellet in an appropriate volume of hypotonic buffer and repeat stages vi and vii. The pelleted chromosomes are ready for transfection and should be used immediately.

Chromosome recovery can be monitored by staining an aliquot of the preparation with Giemsa and counting the number of chromosomes in the aliquot under the microscope. Ten percent of total donor chromosomes are generally recovered during chromosome preparation.

5.3 Chromosome transfer

This part of the procedure is very similar to DMGT. Chromosomes are precipitated onto cells with calcium chloride and several hours later the cells are treated with a membrane-permeabilizing reagent. Again it is important to use plastic pipettes and tubes throughout. The protocol presented follows closely the method recommended by Nelson *et al.* (41).

(i)   The day before the transfection seed 10×9 cm diameter dishes with 5×10^5 cells per dish. The cells must be in a low phosphate medium such as Dulbecco's modified Eagle's medium (DMEM).
(ii)   Resuspend 10^8 chromosomes in 9 ml of transfection buffer.
(iii)   Slowly add 1 ml of 1.25 M $CaCl_2$ to the chromosome suspension, simultaneously bubbling air through the mixture.
(iv)   Allow the calcium phosphate−DNA mix to form for $20-30$ min at room temperature.
(v)   Add 1 ml of the transfection mix to the medium of each 9 cm dish of recipient cells. Incubate the cells and chromosomes for $4-6$ h at 37°C in a humidifying incubator.
(vi)   Remove the medium and add 10 ml of wash buffer.
(vii)   Remove the wash buffer and treat the cells with 1 ml of glycerol shock medium for 4 min at room temperature.
(viii) Rinse the cells three times with wash buffer and incubate overnight at 37°C in non-selective growth medium.
(ix)   After 24 h replace the growth medium with selective medium. Change the medium every $3-4$ days.
(x)   Colonies should appear in $14-21$ days.

5.4 Pre-selection

In DMGT, donor genomic DNA is routinely co-transfected together with plasmids encoding dominantly selectable markers as, by pre-selecting for incorporation of the plasmid DNA, a 1000-fold enrichment for cells containing the cellular gene of interest is obtained. A similar pre-selection strategy can be utilized in CMGT. For co-transformation, an appropriate aliquot of plasmid DNA is added to the chromosome suspension (stage ii) prior to precipitation using calcium chloride (stage iii). We routinely add sufficient plasmid DNA to achieve a 20:1 (w/w) ratio of cellular chromosomal

DNA to plasmid DNA. Selection for the plasmid is applied to the cells 24 h after chromosome transfection.

5.5 Trouble-shooting and variant protocols

Many different procedures have been described for releasing and purifying chromosomes from cells that have been blocked at metaphase. The purification procedures range from those that yield highly purified preparations to those in which the final chromosome suspension is relatively crude and contaminated with other cellular components. We have preferred to use crude preparations for transfection, primarily because they can be prepared in a short space of time, thus minimizing chromosome disintegration.

An analysis of CMGT by Lewis *et al.* (42) has indicated the existence of a linear correlation between donor chromosome dosage and CMGT transfection frequency. Most subsequent CMGT experiments have consequently utilized as many chromosomes as possible for transfection, however there are practical limitations on the amount of chromosomal material that can be handled. A major drawback in using very large numbers of donor cells is the increased viscosity and stickiness of the resulting chromosome suspension, which, by promoting chromosome−chromosome agglutination, can result in the simultaneous transfection of asyntenic chromosomal fragments. For this reason, we have not added more than 20 chromosomes for each recipient cell. Although we have exploited mechanical disruption, chemical treatment using a mild detergent such as digitonin can also release chromosomes (43).

In our experience the transfection procedure is robust and rarely causes any problems providing the recipient cell is transfectable. It may be necessary to optimize the 'shock' conditions by titration of the concentration of glycerol and time of application. Further consideration of calcium phosphate-mediated transfection is presented in Section 6.

CMGT generates recipient cells containing fragments of donor chromosomes: in some cases the fragments integrate into the recipient genome and in some cases the fragments replicate autonomously. It has not been possible to identify the variables that control the sizes of the transferred fragments and in most experiments clones are generated with widely differing amounts of donor material. In all the cases we have analysed in detail the transfused fragments have been rearranged: interstitial deletions are frequent and there is an over-representation of alphoid sequences derived from the centromeric region (12,21). Interstitial deletions have also been reported by others (22).

6. DNA-MEDIATED GENE TRANSFER

6.1 Introduction

A large number of methods have been developed for introducing cloned DNA sequences into mammalian cells. These methods include precipitation by calcium phosphate or DEAE dextran, electroporation, the use of infectious defective viruses and the fusion of prokaryote and yeast protoplasts with mammalian cells. Although several of these approaches can be used with uncloned genomic DNA the only method that has been widely exploited is calcium phosphate-induced precipitation. The precise mechanism of DNA uptake and incorporation by the recipient cells is not understood, however

Table 7. Solutions required for DMGT.

Growth medium	This must be a low phosphate medium such as DMEM.
Selection medium	

2 × concentrated Hepes-buffered saline (2 × HBS) pH 7.1 ± 0.05	The pH is critical and must be checked if the reagent has been stored.
	  50 mM Hepes
	290 mM sodium chloride
	  1.5 mM sodium phosphate (equal amounts of mono and dibasic salts).

Hepes-buffered saline (1 × HBS) pH 7.1	25 mM Hepes 145 mM sodium chloride 0.75 mM sodium phosphate (equal amounts of mono and dibasic salts).
1.25 M calcium chloride	
Shock solution	15% glycerol in 1 × HBS

only a small portion of cells in the recipient culture incorporate DNA. By analogy with bacterial systems these cells have been termed 'competent'. The amount of DNA incorporated is a function of the particular cell line used. Mouse L cells incorporate several million base pairs of exogenous DNA, human fibroblasts incorporate only a fraction of this amount (44). There have been few studies on the maximum size of contiguous DNA that can be transferred, however it is unusual for more than 100 kb to be transferred intact. It is not known if this limitation is due to the recipient cells or difficulties in preparing high molecular weight DNA. Recent advances in the handling and separation of high molecular weight DNA fragments will allow these possibilities to be checked.

6.2 Calcium phosphate-mediated DNA transfections

(i) Prepare the solutions described in *Table 7*.
(ii) Plate the cells in 9 cm diameter Petri dishes (5×10^5 cells/dish) the day before transfection.
(iii) Prepare DNA for transfection by adding 20 μg of DNA to 0.5 ml of 2 × Hepes-buffered saline (HBS) (for one 9.0 cm diameter dish of 5×10^5 cells, larger volumes, sufficient for up to 10 dishes, can be prepared) and adding distilled water to give a volume of 0.9 ml.
(iv) Slowly add 0.1 ml of 1.25 M CaCl$_2$ to the DNA solution whilst gently bubbling air through the mixture. This ensures good mixing and prevents large aggregates forming.
(v) Leave at room temperature for 20–30 min until the solution is opaque and has a slightly 'milky' appearance.
(vi) Add the DNA mix directly to the cells without removing the growth medium, and leave for 4 h at 37°C.
(vii) Rinse the dishes once with 1 × HBS.

(viii) Add 1 ml of glycerol shock solution, leave for 4 min at room temperature.
(ix)   Wash the dishes 2−3 times with 1 × HBS.
(x)   Add fresh non-selective medium.
(xi)   After 24 h remove the non-selective medium and add selective medium. Replace the medium every 3−4 days. Transfectant colonies should appear within 14−21 days.

6.3 Co-transfection and pre-selection

Wigler *et al.* (45) demonstrated that the competent cells in a recipient population incorporate large amounts of donor DNA and that a single recipient cell can incorporate, at a single genomic site, several unrelated donor DNA molecules. These phenomena allow the isolation of the transfected (competent) subpopulation of a mass culture of recipient cells and the 'tagging' of mammalian genes.

If donor genomic DNA is mixed with plasmid DNA encoding a mammalian-selectable marker, selection for the plasmid gene after transfection results in isolation of the whole population of transfected cells. The enrichment gained facilitates the subsequent isolation of recipient cells when only a weak selection or enrichment is available for the donor gene of interest. This approach has been successfully applied to the cloning of genes encoding cell surface antigens using antibody-based enrichment on the FACS (46).

DNA of the selectable plasmid is ligated to donor genomic DNA by the recipient cell prior to integration. This results in 'tagging' of mammalian sequences and can simplify the isolation of donor genes after several rounds of DNA transfer (47).

In co-transfer experiments we have used a mixture of 1 μg of plasmid and 20 μg of genomic DNA. The DNAs are mixed immediately prior to calcium chloride addition.

6.4 Trouble-shooting and variant protocols

Not all cells can be transfected at high frequency with genomic DNA. Some cell lines are refractory to transfection and some cells, such as human fibroblasts, can be transfected with plasmid at a high frequency but will only incorporate small amounts of genomic DNA. Mouse L cells are transfectable at high frequency with genomic DNA and should be used as a positive control in any attempts to transfect a new cell line. It is possible that alternative techniques for direct genomic DNA transfer, such as electroporation or liposome-mediated transfer may extend the range of transfectable cell lines.

With L cells the calcium phosphate precipitation method is robust. Despite folk-lore to the contrary, we have obtained good transfection with precipitates (stage iv) which are either only just opalescent or, at the other extreme, flocculant. Nevertheless, the former condition is probably preferable.

The glycerol shock enhances transfection frequencies by 2- to 5-fold. The shock conditions are cell line specific and both the concentration of glycerol and the time of application should be titrated for each new recipient cell line.

7. CONCLUDING REMARKS

It is hoped that the methods presented in this chapter will serve as a starting point for those wishing to transfer genomic sequences from one mammalian cell to another. The protocols are not sacrosanct and the procedures may contain elements closer to magic than science. If a protocol fails to work; first, test the cloning efficiency of the recipient cell under the conditions of the experiment; second, consult a laboratory where the protocol works; third, titrate the variables. Good luck.

8. ACKNOWLEDGEMENTS

P.N.Goodfellow would like to thank Dr V.Van Heyningen for teaching him to grow and fuse cells. We gratefully acknowledge the efforts of all our colleagues, past and present, from the Laboratory of Human Molecular Genetics, who have struggled to make hybrids and transfectants. Editorial assistance with this manuscript was expertly provided by Mrs C.Middlemiss.

9. REFERENCES

1. Barski,G., Sorieul,S. and Cornefert,F. (1960) *C.R.Acad. Sci.*, **251**, 1825.
2. Ruddle,F.H. (1972) *Adv. Hum. Genet.*, **3**, 173.
3. Kohler,G. and Milstein,C. (1975) *Nature*, **256**, 495.
4. Weiss,M. and Green,H. (1967) *Proc. Natl. Acad. Sci. USA*, **58**, 1103.
5. Human Gene Mapping 9 (1988) *Cytogenet. Cell Genet.*, **46**, 1.
6. Bengtsson,B.O., Nabholz,M., Kennett,R., Bodmer,W.F., Povey,S. and Swallow,D. (1975) *Somat. Cell. Genet.*, **1**, 41.
7. Fournier,R.E.K. and Ruddle,F.H. (1977) *Proc. Natl. Acad. Sci. USA*, **74**, 219.
8. Petit,C., Levilliers,J., Ohm,C. and Weiss,M. (1986) *Proc. Natl. Acad. Sci. USA*, **83**, 2561.
9. Weissman,B.E., Saxon,P.J., Pasquale,S.P., Jones,G.R., Geiser,A.G. and Stanbridge,E.J. (1987) *Science*, **236**, 175.
10. McBride,O.W. and Ozer,H.L. (1973) *Proc. Natl. Acad. Sci. USA*, **70**, 1258.
11. Goss,S.J. and Harris,H. (1975) *Nature*, **255**, 680.
12. Pritchard,C.A. and Goodfellow,P.N. (1987) *Genet. Dev.*, **1**, 172.
13. Graham,F.L. and van der Eb,A.J. (1973) *Virology*, **52**, 456.
14. Neumann,E.M., Schafer-Ridder,M., Wang,Y. and Hofschneider,P.H. (1982) *EMBO J.*, **1**, 841.
15. Mann,R., Mulligan,R.C. and Baltimore,D. (1983) *Cell*, **33**, 153.
16. Capecchi,M.R. (1980) *Cell*, **22**, 479.
17. Tunnacliffe,A., Parkar,M., Povey,S., Bengtsson,B.O., Stanley,K., Solomon,E. and Goodfellow,P. (1983) *EMBO J.*, **2**, 1577.
18. Lowy,I., Pellicer,A., Jackson,J.F., Sim,G.K., Silverstein,S. and Axel,R. (1980) *Cell*, **22**, 817.
19. de Jonge,A.J.R. and Bootsma,D (1984) *Int. Rev. Cytol.*, **92**, 133.
20. Fournier,R.E.K. (1981) *Proc. Natl. Acad. Sci. USA*, **78**, 6349.
21. Goodfellow,P.N. and Pritchard,C.A. (1988) *Cancer Surveys*, in press.
22. Porteous,D.J. (1987) *Trends Genet.*, **3**, 171.
23. Ringertz,N.R. and Savage,R.E. (1976) *Cell Hybrids*. Academic Press, New York.
24. Shay,J.W. (1982) *Techniques in Somatic Cell Genetics*. Plenum Press, New York.
25. Szybalski,W., Szybalska,E.H. and Ragni,G. (1962) *Natl. Cancer Inst. Mongr.*, **7A**, 75.
26. Nabholz,M., Miggiano,V. and Bodmer,W.F. (1969) *Nature*, **223**, 358.
27. Puck,T.T. and Kao,F.T. (1982) *Annu. Rev. Genet.*, **16**, 225.
28. Mulligan,R.C. and Berg,P. (1980) *Science*, **209**, 1422.
29. Southern,P.J. and Berg,P. (1982) *J. Mol. Appl. Genet.*, **1**, 327.
30. Smithies,O., Gregg,R.G., Boggs,S.S., Koralewski,M.A. and Kucherlapati,R.S. (1985) *Nature*, **317**, 230.
31. Tunnacliffe,A., Jones,C. and Goodfellow,P.N. (1983) *Immunol. Today*, **4**, 230.
32. Harris,H. and Watkins,J. (1965) *Nature*, **205**, 640.
33. Pontecorvo,G. (1975) *Somat. Cell. Genet.*, **1**, 397.
34. Schneiderman,S., Farber,J.L. and Baserga,R. (1979) *Somat. Cell. Genet.*, **51**, 263.
35. Zorn,G.A., Lucas,J.L. and Kates,J.R. (1979) *Cell*, **18**, 659.
36. McNeill,C.A. and Brown,R.L. (1980) *Proc. Natl. Acad. Sci. USA*, **77**, 5394.
37. Goodfellow,P.N., Banting,G., Trowsdale,J., Chambers,S. and Solomon,E. (1982) *Proc. Natl. Acad. Sci. USA*, **79**, 1190.
38. Mullinger,A.M. and Johnson,R.T. (1976) *J. Cell. Sci.*, **22**, 243.

39. Wigler,M.H. and Weinstein,I.B. (1975) *Biochem. Biophys. Res. Commun.*, **63**, 669.
40. Prescott,D.M., Myerson,D. and Wallace,J. (1972) *Exp. Cell Res.*, **71**, 480.
41. Nelson,D.L., Weis,J.H., Przyborski,M.J., Mulligan,R.C., Seidman,J.G. and Houseman,D.E. (1984) *J. Mol. Appl. Genet.*, **2**, 563.
42. Lewis,W.H., Srinivasan,W.H., Stokoe,N. and Siminovitch,L. (1980) *Somat. Cell. Genet.*, **6**, 333.
43. Porteous,D.J, Boyd,P.A., Christie,S., Cranston,G., Fletcher,J.M., Gosden,J.R., van Heyningen,V., Rout,D., Seawright,A., Simola,K.O.J. and Hastie,N. (1987) *Proc. Natl. Acad. Sci. USA*, **84**, 5355.
44. Hoeijmakers,J.H.J., Odijk,H. and Westerveld,A. (1987) *Exp. Cell Res.*, **169**, 11.
45. Wigler,M., Sweet,R., Sim,G.-K., Wold,B., Pellicer,A., Lacy,E., Maniatis,T., Silverstein,S. and Axel, R. (1979) *Cell*, **16**, 777.
46. Kuhn,L.C., McClelland,A. and Ruddle,F.H. (1984) *Cell*, **37**, 95.
47. Westerveld,A., Hoeijmakers,J.H.J., van Duin,M., de Wit,J., Odijk,H., Pastin,K.A., Wood,R. and Bootsma,D. (1984) *Nature*, **310**, 425.

CHAPTER 2

Genome mapping by restriction fingerprinting

ALAN COULSON and JOHN SULSTON

1. INTRODUCTION

Genome mapping is a rapidly evolving field. This is both good news and bad news for the authors of an article on the subject. On the one hand, it means that we do not have to be very selective, because we are not yet sure what is important and what is not. On the other hand, we are unable to come to any conclusion about the ideal approach to creating a genome map. Indeed, because of the technical advances since we began this project, we would be unlikely to choose exactly the same approach if we were starting it again today.

First, a definition. Genome mapping refers to the generation of an ordered clone library that fully represents a genome (or a defined part of a genome), together with enough genetic landmarks to allow accurate alignment of the physical and genetic maps. As a mapping project proceeds, this alignment becomes better and better until the two maps are synonymous. A genome map need not necessarily include a restriction map, though some techniques automatically generate one as a by-product.

The principal purpose of mapping is to facilitate both the cloning of known genes and the genetic placing of known clones. A map also provides for the archiving of all information about the structure of a genome, and thus allows easy communication between researchers regarding the disposition of their clones.

2. STRATEGY

2.1 Comparison of clones

Central to any genome mapping project is a procedure for comparing clones with one another in order to detect overlaps between them. Most simply, overlaps can be detected directly by hybridization. However, hybridization between clones is not in itself a satisfactory criterion for matching, because false positives can arise from dispersed repeats, particularly in eukaryote genomes. Furthermore, reliance on direct pairwise comparison is an encumbrance for large projects, where large numbers of clones must be considered. It is more satisfactory to find a procedure in which data can be collected from each clone in turn and stored for analysis at later times.

For these reasons, most strategies for genome mapping rely upon digestion with one or more restriction enzymes followed by measurement of the size of the resulting fragments, and a variety of protocols have been devised to this end. In this account we shall be primarily concerned with the approach that we have taken to map the genome of the nematode *Caenorhabditis elegans* (80×10^6 bp) (1). The matching procedure used for *C. elegans* has been termed 'fingerprinting', because it is based on an unordered and incomplete set of fragments that are characteristic, but not fully descriptive, of the clone; the fragments are then separated by thin polyacrylamide gel electrophoresis. Protocols are also given for two other, closely related, fingerprinting procedures (Gibson; Brownlee and Knott).

2.2 Vectors

In order for a strategy based upon the analysis of randomly selected clones to proceed efficiently, it is clearly important that the libraries used should be as random as possible. It is also desirable that the cloned inserts should be as large as possible, consistent with their subsequent utility for genetic biochemistry.

For our initial efforts, we chose to use cosmids because of the relatively large amount of DNA (40 kb) that they can accommodate. Since the original cosmid vectors were introduced some 10 years ago by Collins and Hohn (2), a considerable number of vectors have been constructed. Those used in the *Caenorhabditis* project have been pJB8 (3) and the Lorist series, in particular Lorist2 and Lorist6 (4,5) derived from LoristB of Little and Cross (6). These latter have features which give rise to a constant copy number, and, perhaps more importantly for the techniques described here, elements which should enhance the randomness of libraries, such as transcriptional terminators to prevent interference of vector genes by transcription from the cloned inserts. They also contain SP6 and T7 phage promoter sequences to facilitate directed walking. Protocols for preparation, cleavage and dephosphorylation of vector DNA are described by Maniatis *et al.* (7).

λ2001 (8) is the only lambda vector to have been used in the *Caenorhabditis* project. Lambda inserts are only half the size of cosmid inserts, but the clones are more stable and, at least in some areas, the representation of the genome in lambda banks is superior (A. Spence, personal communication).

The recent introduction of a vector for generating yeast artificial chromosomes (YACs) (9) promises to revolutionize mapping strategies. By this means, fragments of DNA at least an order of magnitude larger than cosmid inserts can be cloned, and probably banks more fully representative of the genome can be generated. The fingerprinting method that we have adopted is suitable only for the smaller YACs, but we are using hybridization to link the larger ones to the cosmids (see Section 3.4). Future mapping strategies are likely to involve matching techniques specifically devised for YACs.

2.3 Selection of clones

Once a clone bank has been constructed and a matching technique has been devised, there remains one further option — the manner in which clones are to be selected for analysis. For all but the smallest projects, it is most efficient to pick clones randomly at first. Later, when this approach becomes unproductive, the clones needed to fill the remaining gaps in the map must be found by hybridization or by other means.

3. PROCEDURES

Recipes for buffers and generally useful reagents will be found in Section 5.

3.1 **Library construction**

3.1.1 *Cosmids*

Because cosmid libraries have been our principal working material, we describe their preparation in some detail.

(i) *Isolation of genomic DNA.* The example given here is for DNA extraction from nematodes, but is generally applicable. The essence is minimal manipulation.

Worms grown in liquid culture (10) are washed, quick frozen and stored in 1 g aliquots in liquid N_2.

(a)   For each preparation, grind one aliquot to a powder in a mortar cooled in liquid N_2, then mix very gently, in portions, into 30 ml of 100 mM ethylenediamine-tetraacetic acid (EDTA) pH 8.0, 5 mM Tris−HCl pH 8.0, 0.5% sodium dodecyl sulphate (SDS), 50 μg/ml proteinase K.

(b)   Incubate at 50°C for 2 h.

(c)   Cool in ice, add an equal volume of phenol and extract by slow rolling at 4°C for 15 min. Centrifuge and carefully remove the aqueous layer with a wide mouth pipette.

(d)   Add two volumes of 95% ethanol, and gently spool the DNA around a wide rod. Wash three times with 70% ethanol, air dry briefly, and disperse in 3 ml of 10 mM Tris−HCl pH 7.4, 0.1 mM EDTA (TE).

CsCl gradient purification is not necessary for the efficient production of cosmids.

(ii) *Preparation of fragments.* In order for a map compiled by a fingerprint matching method to be free from error, it is important that cloned fragments should not be derived by ligation of unconnected fragments. We therefore employ a two-gel purification procedure, which in our hands gives a much cleaner size fractionation than sucrose density gradients and a satisfactory yield of fragments clonable at high efficiency.

When considering the particular partial digest, it is as well to integrate the library with the fingerprinting system to be used, in order to prevent the complication of 'end effect' bands in the fingerprint patterns. Thus, for instance, a partial *Sau*3AI library fingerprinted with *Hind*III and *Sau*3AI will give artefact-free data. With cosmid clones, where a fingerprint may consist of 20−30 bands, this is not an absolute requirement but end effects are a considerable inconvenience for the mapping of lambda clones by this method.

(a)   Establish suitable partial digestion conditions [see for example Maniatis *et al.* (7)] and select four of increasing severity yielding the fragment size range required (i.e. ~40 kb).

(b)   Digest four 40 μg aliquots of freshly extracted DNA accordingly. Freeze the digests.

(c)   Prior to running a preparative gel, test small samples (~50 ng) alongside λ *Hind*III markers on a 150 ml 0.3% HGT TAE (see Section 6) agarose gel (17×17^2), run at about 1 V/cm for 16 h. All subsequent analyses should be done in this way to obtain an accurate estimation of fragment sizes.

(d)   When the DNA has been adequately digested, combine the aliquots, phenol extract and ethanol precipitate.

(e)   Re-disperse in 180 μl of TE. Add 60 μl of non-denaturing dye mix and load into 16 0.5 cm slots of a 250 ml 0.4% LGT TAE agarose gel. (When loading large amounts of partially digested DNA, be careful not to pull a thread of DNA to the surface of the buffer when withdrawing the loading pipette — you may find your DNA unspooling itself into an irrecoverable layer on the buffer surface.)

(f)   Also load λ *Hin*dIII and intact λ markers — on a preparative gel these will not give an accurate size indication, but will aid straight cutting of gel bands. Again, run at 1 V/cm for 20 h and stain with ethidium bromide. (This can be included in the buffer at a concentration of 0.2 μg/ml.)

(g)   Visualize the result with a hand-held long-wave UV lamp (avoid all exposure of the DNA to short-wave UV) and cut 2 mm wide slices through all 16 lanes from apparently undigested material downwards.

(h)   Melt the most likely slices at 68°C for 10 min in Falcon 2063 tubes. Phenol extract, concentrate with isobutanol to 0.3 ml, ethanol precipitate and analyse the fragments as above.

(i)   Re-purify the required size fractions on a second preparative LGT gel, extract the appropriate regions of the bands and finally analyse on a minimally loaded, slowly run analytical gel. You should expect to recover about $2-4$ μg of appropriately sized DNA from 160 μg of starting material.

(j)   Disperse in a final volume of 20 μl TE.

(iii) *Packaging mixes.* There are a number of sources of commercially prepared packaging mixes, but for reasons of economy it may be desirable to make your own. We have used the *Escherichia coli* strains NS428 and NS433 (11) with considerable success. Preparation is very easy and they have yielded up to 10^6 cosmid recombinants per μg of inserted DNA.

There is some evidence that *Eco*K activity in packaging mixes may bias the representation of λ libraries (12), but we have no direct evidence that this is the case in cosmid libraries. *Eco*K$^-$ packaging mixes are available commercially (e.g. Gigapack).

(iv) *Ligation of fragments into vector.* The recipe given is for the ligation of partial *Sau*3AI fragments into Lorist2.

(a)   Add together 1 μl (0.2 μg) of *Bam*HI-cut phosphatased Lorist2; 2 μl ($\sim$0.3 μg) of sized partial *Sau*3AI fragments; 2 μl of 10 $\times$ ligation buffer; 2 μl of 10 mM ATP pH 7.4; 2 μl of 0.1 M dithiothreitol; 10 μl of water and 1 μl of ligase (Boehringer, 1 Weiss Unit). For maximum ligation, seal in a capillary.

(b)   Incubate for 16 h at 14°C.

(c)   Package directly, following the appropriate protocol.

(d)   Store the library at 4°C over chloroform.

(e)   Assay as follows. Dilute 5 μl of library in 0.1 ml of λ dil. Add 0.2 ml of saturated host cells. [We generally use 1046 (13). This should be regularly colony purified, and isolates tested for lack of *rec*A function, by UV sensitivity.] Incubate at 37°C for 20 min. Add 0.5 ml of CY medium, and incubate for a further 50 min. Plate varying volumes on 50 μg/ml kanamycin plates.

You should expect 10^5-10^6 recombinants per microgram of insert DNA.

3.1.2 *Lambdas*

Lambda libraries are constructed like cosmid libraries, but from DNA cut and sized to 15−20 kb. For convenience and efficiency, we separate the λ 2001 vector arms from the stuffer fragment, so that the resulting clones can be plated directly on Q358. Using 'Gigapack' packaging mix, we obtain 5×10^6 recombinants/µg insert DNA.

3.1.3 *Yeast artificial chromosomes (YACs)*

This method was contributed by Robert Waterston. Prepare DNA by the method of Olson *et al.* (23), as modified by Carle and Olson (14). In the case of *C. elegans*, gently lyse young animals, of the endonuclease-deficient strain *nuc-1* (e1392), by prolonged proteinease digestion. Extract with phenol−chloroform after the sucrose gradient. Ligate and transform according to Burke *et al.* (9), but size fractionate over a sucrose gradient after the ligation step.

3.2 **Characterization of clones**

3.2.1 *Preparation of DNA from cosmids*

We use two procedures, both based on the alkaline lysis method of Birnboim and Doly (15); a 2 ml 'miniprep' extraction, yielding 1−3 µg of cosmid DNA and a 250 µl 'microprep' utilizing standard 96-well microtitre plates.

Up to 96 samples at a time can comfortably be worked up by the miniprep procedure using Eppendorf tubes and a rack-holding centrifuge, such as the Eppendorf 5413. Two or four times this number can be extracted by the microtitre plate method using the 'Micronic' system (Flow Laboratories) of 1 ml tubes in a microtitre plate array for growth and storage, in conjunction with a multichannel pipette (Titertek 8-channel) and round-bottom microtitre plates (Corning) for DNA extraction.

(i) *Cosmid DNA 'miniprep' procedure*

(a) Inoculate, by toothpick, 4 ml of $2 \times$ TY + 50 µg/ml kanamycin (or 120 µg/ml ampicillin) in a 15 ml culture tube with a generous portion (∼2 µl) of a streaked colony.

(b) Grow for 16 h at 37°C with shaking.

(c) For storage of aliquots, remove 0.75 ml into screwtop 2 ml ampoules containing 0.75 ml of sterile 50% glycerol. Vortex and freeze immediately at −70°C. Quick freezing ensures that the bacteria remain suspended, and that samples can be recovered by scraping from the surface rather than needing undesirable thawing.

(d) For extraction of cosmid DNA, remove 2 ml of culture into 2.2 ml Eppendorf tubes. Centrifuge for 45 sec (Eppendorf rack centrifuge). Remove supernatants by aspiration.

(e) Resuspend the pellets by vortexing for 15 sec in 250 µl of 50 mM glucose, 10 mM EDTA, 25 mM Tris−HCl pH 8.0.

(f) Leave to stand for 5 min in ice water.

(g) Add 250 µl of 0.2 N NaOH, 1% SDS; mix using ∼15 inversions (do not vortex).

(h) Leave to stand for 5 min in ice water.

(i) Add 200 µl of 3 M sodium acetate pH 4.8; mix using ∼15 inversions (do not vortex).

23

(j)   Leave to stand for 60 min at 0°C.

(k)   Invert the tubes gently a few times; centrifuge for 4 min at room temperature.

(l)   Transfer the supernatants to 1.5 ml Eppendorf tubes; add approximately 0.9 ml of cold 95% ethanol and mix.

(m)   Leave to stand for 20 min at −20°C.

(n)   Centrifuge for 2 min, decant the ethanol and drain briefly.

(o)   Resuspend the pellets in 200 μl of 0.3 M sodium acetate pH 7, 1 mM EDTA.

(p)   Vortex, leave to stand for 15 min at room temperature, then vortex again. Add 400 μl ethanol and mix.

(q)   Leave to stand for 20 min at −20°C and then centrifuge for 2 min.

(r)   Decant, wash the pellets in 95% ethanol, drain for 5 min, dry for 5 min under vacuum and resuspend in 30 μl of TE.

The yield should be 1−3 μg of cosmid DNA.

(ii) *Cosmid DNA 'microprep' procedure* (16).

(a)   Autoclave Micronic tubes (labelled as necessary) in their boxes.

(b)   Dispense 0.6 ml of 2 × TY + 50 μg/ml kanamycin (or 120 μg/ml ampicillin) for each tube (this can conveniently be done using a BCL8000 dispenser with a 5 ml glass pipette on the end of the supplied 6 ml syringe, secured by a short length of rubber tubing).

(c)   Transfer the colonies by toothpick into the tubes. (Leave the toothpicks in the tubes until the box is completed, then remove.)

(d)   Replace the box lid, and fasten into an orbital incubator, preferably arranging the boxes at a tilt to aid aeration and suspension of the cells. Shake at 37°C and 300 r.p.m. for 12−16 h.

(e)   To make glycerol stocks for long-term storage, add 0.3 ml of 50% glycerol to the residue after removing aliquots for DNA extraction. Seal the tubes with Micronic cap strips.

(f)   Use a multichannel pipette (e.g. Titertek) for all subsequent transfers. Transfer 250 μl of cultures to the microtitre plate wells.

(g)   Support the plates on polystyrene foam pads and pellet the cells at 2500 r.p.m. for 2 min (e.g. Centra 4X or 7R centrifuge).

(h)   Discard the medium and drain in an inverted position briefly then soften the pellet by 20 sec of vigorous vortexing on a multi-tube vortexer (e.g. SMI model 2601).

(i)   Add 25 μl of 50 mM glucose, 10 mM EDTA, 25 mM Tris−HCl pH 8.0. Rock the plates to suspend the cells.

(j)   Add 25 μl of 0.2 M NaOH 1% SDS and mix by rocking. Leave to stand at room temperature for 5 min (until the samples are largely clarified).

(k)   Add 25 μl of 3 M sodium acetate pH 4.8. Seal with a Micronic cap-mat. Mix gently by inversion, then shake with increasing vigour.

(l)   Leave to stand for 5 min, then vortex gently for 1 min and centrifuge at 3000 r.p.m. for 5 min.

(m)   Transfer 70 μl of supernatant to a new plate containing 100 μl of isopropanol per well, then rinse the cap-mat, replace and shake the plate and place at −20°C for 30−60 min.

(n)   Centrifuge at 3000 r.p.m. for 5 min, discard the supernatant and drain for 5 min on successive tissues, finishing with a sharp tap.

(o)   Add 25 μl of water and resuspend the pellets. Add 25 μl of 4.4 M LiCl, replace the cap-mat, shake and place at 4°C for at least 1 h.

(p)   Centrifuge at 3000 r.p.m. for 5 min, and transfer 50 μl of supernatant, tilting the plate, to a new plate containing 100 μl of isopropanol per well.

(q)   Replace the cap-mat, shake and place at -20°C for 1 h then spin at 3000 r.p.m. for 5 min, discard the supernatant and drain well.

(r)   Add 200 μl of 95% ethanol, centrifuge and drain extremely well (>10 min), ending with a sharp tap on a tissue.

(s)   Dry in a vacuum desiccator and resuspend the pellets in $5-20$ μl of TE. Seal the wells with parafilm and store at -20°C.

The following refinements may be helpful under some circumstances:

(1)   leaving the cells, in the plate, for 2 h or so at $+4$°C before the first centrifugation;

(2)   use of non-tissue culture-treated microtitre plates for the second and third stages;

(3)   a second ethanol wash at the end.

3.2.2 *Preparation of DNA from lambda clones*

This convenient and reliable method was devised by Toby Gibson at the Laboratory of Molecular Biology.

(i)   Grow overnight phage streaks on λ plates layered with 50 μl of saturated cells in 4 ml of top agar.

(ii)   Scrape up each streak with a toothpick into 2 ml of CY + 10 mM $MgCl_2$ + 1/200 dilution of saturated cells.

(iii)   Shake at 37°C until lysed (typically 10 h).

(iv)   Centrifuge for 5 min in 2 ml Eppendorf tubes.

(v)   Pour the supernatant into a new 2 ml tube containing 2 μl of DNase/RNase (20 mg/ml each).

(vi)   Leave to stand at room temperature for 30 min or longer.

(vii)   Add 400 μl of 25% polyethylene glycol (PEG) in 3 M NaCl. Shake, then leave at 4°C for 30 min to overnight.

(viii) Centrifuge for 5 min and discard the supernatant. Re-spin and aspirate the remaining supernatant.

(ix)   Add 200 μl of 10 mM EDTA, 10 mM Tris$-$acetate pH 8; vortex, then add 100 μl of phenol. Vortex, leave on ice for 5 min, vortex and centrifuge for 15 sec.

(x)   Transfer the aqueous phase to 1.5 ml Eppendorf tubes containing 20 μl of 3 M sodium acetate, add 200 μl of isopropanol, vortex and leave to stand at -20°C for at least 30 min.

(xi)   Spin for 5 min, discard the supernatant and drain, wash with 95% ethanol then spin for 5 min, drain on paper, dry under vacuum and resuspend in 30 μl of TE.

3.2.3 *Fingerprinting*

The essential feature of the fingerprinting reaction is that it should produce, on average, a suitable number of labelled fragments, separable on a high resolution polyacrylamide

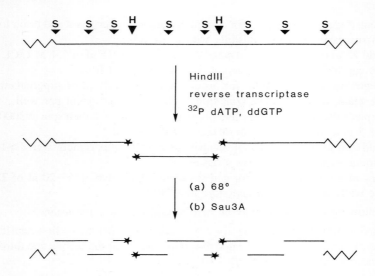

Figure 1. Fingerprinting by the double cutting method. Zigzag line: vector; straight line: insert; H: *Hind*III site; S: *Sau*3AI site. Modified from Coulson *et al.* (1).

gel (40−2000 bp) such that, generally, cosmids that overlap to the extent of 1/2−1/3 of their length will be unambiguously recognized as matching. A number of schemes may be envisaged to achieve this end.

(i) *Double cutting method.* The fingerprinting protocol used for the *C. elegans* mapping project was devised by Sydney Brenner and Jonathan Karn (*Figure 1*). It employs a *Hind*III digest, end labelling and a *Sau*3AI digest, and yields, on average, about 23 bands from a cosmid of 36% GC. Such a scheme can be easily modified to suit the base composition of the particular genome under study (see Section 4.2).

(a)   Aliquot 2 μl (~50−100 ng) of DNA into a 96-well round-bottom (non-tissue culture-treated) microtitre plate on ice. (Use a Titertek multichannel pipette for transfer from 'microprep' microtitre plate.)

(b)   Using a Hamilton PB600-1 repetitive dispenser fitted with a disposable tip, aliquot 2 μl of the following mix onto the side of each well:

   (1)   Dry 40 μl (40 μCi) of 400 Ci/mmol [^{32}P]dATP (ethanolic) [or add 8 μl (80 μC) of 800 Ci/mmol aqueous [^{32}P]dATP].

   (2)   Add 160 μl of water, 40 μl of 10 × medium salt buffer, 4 μl of 10 mg/ml RNase (Section 5), 10 μl of 0.5 mM ddGTP (to drive the fill-in reaction to completion), 4 μl (40 units) of *Hind*III, and 4 μl (40 units) of avian myeloblastosis virus (AMV) reverse transcriptase.

(c)   Spin briefly (up to 2000 r.p.m. and down) then seal the plate (the air space can be reduced by pushing a trimmed microtitre plate into the wells of the reaction plate). Seal around the edges with parafilm.

(d)   Incubate for 45 min at 37°C and then for 25 min at 68°C (to kill the reverse transcriptase). Cool on ice before opening.

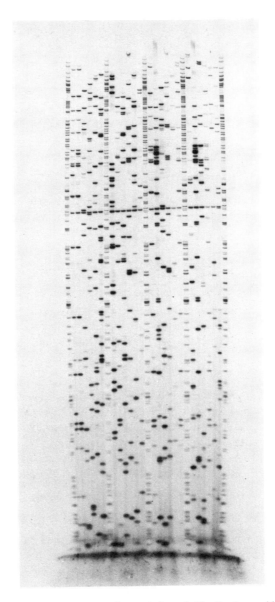

Figure 2. Autoradiogram of a *Hind*III/*Sau*3AI fingerprinting gel. The five lanes with closely spaced bands are markers. The band present in all the sample lanes is derived from the vector. Several overlaps can be seen, because the clones were not picked at random but were pre-selected by hybridization to a mixture of probes.

(e) Aliquot 2 µl of the following mix onto the side of each well: 200 µl of water; 20 µl of 10 × medium salt buffer; approximately 80 units of *Sau*3AI (this should be >30 units/µl, to prevent miscutting due to excess glycerol).

(f) Spin briefly and reseal. Incubate for 2 h at 37°C.

(g)   Add 4 µl of formamide/dye/EDTA to each well (Section 5) and denature at 80°C for 10 min before gel loading (*Figure 2*).

(ii) *Single cutting methods using HinfI*. Method 1, contributed by George Brownlee and Vroni Knott (24).

(a)   To each well of a 96-well microtitre plate, on ice, add 4 µl of digestion mix (10 mM Tris−HCl pH 7.4, 10 mM $MgCl_2$, 50 mM NaCl, 250 units/ml of *HinfI*).

(b)   Add 1 µl of DNA (4−20 ng).

(c)   Seal the wells (plate sealer, Titertek, Flow Laboratories) and incubate for 60 min at 37°C.

(d)   To each well add 2 µl of labelling mix (10 mM Tris−HCl pH 7.4, 10 mM $MgCl_2$, 6 mM dithiothreitol, 0.1 mM each of dATP, dGTP and dTTP, 0.2−0.4 µCi [^{32}P]dCTP (3000 Ci/mmol), 0.1 unit of Klenow).

(e)   Incubate for 10 min at room temperature and add 7 µl of formamide dye mix.

(f)   Incubate, unsealed, for 20 min at 80°C before gel loading.

(g)   Carry out electrophoresis as described below, but use 6% polyacrylamide (38:2). After autoradiography, digitize the bands between 25 and 220 bp in length.

This procedure generates simpler output data than Method 2, because only a quarter of the sites are labelled and only the lower molecular weight bands are digitized. Thus the method is suited to inserts in the size range 20−200 kb, and manual entry for cosmid clones is feasible.

Method 2, contributed by Toby Gibson.

(a)   Place 2 µl of DNA (0.1−0.2 µg) in each well of a 96-well microtitre plate, add 2 µl of digestion mix (3 units/µl *HinfI*, 0.2 µg/µl RNase, 20 mM Tris−HCl pH 8, 7 mM $MgCl_2$, 10 mM dithiothreitol) to the side of each well.

(b)   Centrifuge briefly (1000 r.p.m.), seal with parafilm and incubate at 37°C for 1 h.

(c)   Dispense 2 µl of labelling mix (0.2 units/µl reverse transcriptase, 80 nCi/µl [^{35}S]thiodATP, 40 µM each of dCTP, dGTP, dTPP) and incubate for 15 min at 37°C.

(d)   Dispense 4 µl of formamide dye mix.

(e)   Denature at 80°C for 20 min before gel loading (*Figure 3*).

This procedure generates several times as many bands as double-cutting does, and a scanner is therefore essential for satisfactory data input. It is well suited to inserts in the size range 5−50 kb. The insert size can be estimated from the sum of band sizes (complicated by superimposed bands), but, since *HinfI* is not used for making clone banks, anomalous junction fragments are present.

3.2.4 *Polyacrylamide gel electrophoresis*

Fractionation of the fingerprint fragments is on denaturing gels (which give better resolution than non-denaturing gels) essentially as used for DNA sequencing, with some precautions taken to reduce distortion. The gel is bonded to one plate by pre-treatment of the plate with methacryloxypropyltrimethoxysilane (17), to prevent distortion of the 4% gel prior to loading the samples, and a metal plate (3 mm thick aluminium) is

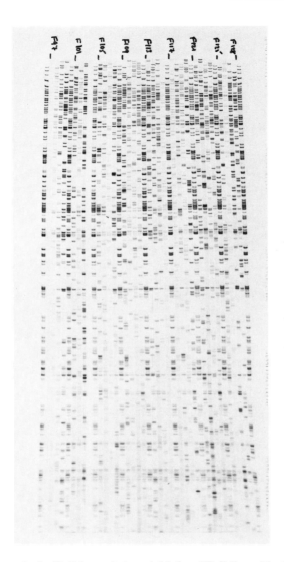

Figure 3. Autoradiograph of a *Hin*fI fingerprinting gel. Markers (*Hin*fI-digested lambda DNA) are placed in **lane 1** and subsequently every fifth lane. Common bands in all sample lanes are vector. Clones derive from a library prepared from partial *Fnu*DII-digested *E. coli* DNA, blunt-end inserted at the *Sca*I site of Lorist6. Clones were prepared from 0.2 ml of culture medium (see Section 3.2.1) and 1/7th of the DNA sample was used. Film exposure was for 2 days.

clamped to the gel during electrophoresis to conduct heat evenly to reduce 'smiling'. Gels are prepared as follows.

(i)   Add 5 μl of methacryloxypropyltrimethoxysilane and 50 μl of 10% acetic acid to 3 ml of ethanol and wipe a tissue moistened with this solution once over a well cleaned 20 × 40 cm unnotched gel plate.

(ii)   Rinse very thoroughly with ethanol after brief drying.

(iii)   Treat the notched gel plate with 'Repelcote' silane, and tape the plates, separated by 0.35 mm spacers, to make the gel template.

(iv)   Take 40 ml of gel mix, add 40 μl of TEMED, 0.35 ml of 10% ammonium persulphate (contrary to popular belief, it is not necessary to make this solution fresh or even frequently) and pour carefully.

(v)   Clamp the template and position the well-former (cut to give 2 mm slots separated by 1 mm). Leave to set for at least 30 min.

(vi)   Denature the samples at 80°C for 10 min, along with, for each gel, 3 μl of marker digest (see below).

(vii) Load about 3 μl of each sample, interspersed every six lanes with 1 μl of marker.

(viii) Electrophorese for 1.75 h at 30 W (limiting current), until the bromophenol blue dye is about 1 cm from the bottom of the gel.

(ix)   Separate the plates (the gel should adhere firmly to the acrylosilane-treated plate — if this treatment has been over-liberal it may adhere to both) and fix the gel in 10% acetic acid for 15 min; wash in tap water for 30 min and dry on an 80°C hot plate ($\sim$30 min). Autoradiography for 2 days, without an intensifying screen, should be sufficient.

(x)   Remove dried gel from the plates by immersion in a bucket of 20% 'Decon' for 2 days. The plates are simply rinsed under a tap prior to re-use.

The marker digest for fingerprinting gels is made up by the following procedure.

(i)   Mix 35 μl of water; 5 μl of 10 × medium salt buffer; 2.5 μl of λ S7 DNA; 10 units of *Sau*3AI.

(ii)   Incubate at 37°C for 60 min.

(iii) Add 4 μl of 800 Ci/mmol [^{35}S]dATP; 2 μl of 10 mM dGTP; 2.5 μl of 10 mM ddTTP; 10 units of AMV reverse transcriptase and incubate at 37°C for 30 min.

3.3 Computing procedures

The organization and use of the software will be described only briefly here: a more complete account has been given by Sulston *et al.* (18).

3.3.1 *Data entry*

Data entry can be either manual (using a digitizer such as a Grafbar) or semi-automatic (using a scanning densitometer and image processing system).

(i) *Manual.* At the beginning of a project, prepare a grid with lines drawn widthways at the positions of the principal marker bands of a representative film. At the beginning of each data entry session, place the grid on the digitizing tablet, and enter its position by means of fiducial marks. Lay the autoradiogram on the grid, and enter the positions of the sample bands in turn, locally aligning the markers with the grid lines as digitization proceeds.

(ii) *Scanning and image processing.* To scan a film, activate the image processing package in the host computer and feed the film in the appropriate orientation into the densitometer. In our system, scanning takes 3.5 min and image processing of a 17 × 40 cm film takes a further 2 min of Vax c.p.u. time. An AED workstation is next used to assess the image processing and to edit the data. Run the picture display

```
1000-F59C6    ( 19b,    0)
                              8 matches    888-ZK227    ( 10b, 545)   0.5E-07        0    11d   C27B11
                             10 matches    127-C27B11   ( 23b, 545)   0.2E-05        0     2    canon
                              8 matches     59-C06D11   ( 18b, 545)   0.3E-04        1     1
                              8 matches    470-T22G2    ( 20b, 545)   0.6E-04        0     1    C27B11
                              8 matches    663-F25F11   ( 21b, 732)   0.9E-04        0     3    K04H8
                difmap:       7 matches    211-AD4      ( 25b, 755)   0.2E-02        0     8    canon
                difmap:       7 matches    217-BC8      ( 24b, 755)   0.1E-02        0     7    AD4
                difmap:       8 matches    272-R03F7    ( 29b, 755)   0.8E-03        0     7    canon
                difmap:       6 matches    473-T23E7    ( 20b, 803)   0.3E-02       11     6    AF2
                difmap:       6 matches    659-F23F6    ( 21b, 680)   0.4E-02        5     6    C25E3

1000-F59C7    ( 18b,    0)
                             16 matches    540-E01F12   ( 20b, 483)   0.5E-15      109    2d    CE8
                             14 matches    670-F27D9    ( 19b, 483)   0.7E-12      109     1    CE8
                             10 matches    216-BF1      ( 13b, 483)   0.1E-08      109     1    CE8
                             10 matches    777-F53F9    ( 13b, 483)   0.1E-08      109     0    CE8
                              9 matches    308-C51E9    ( 13b, 483)   0.3E-07      109     1    CE8
                             11 matches   1000-F59C3    ( 19b,   0)   0.1E-07        0     1
                             10 matches    706-F36D1    ( 26b, 714)   0.3E-05        9     1    canon
```

Figure 4. Output from MAPSUB. Header of each section, at left, describes incoming clone in standard format: gel number, clone name, number of bands, contig number. Subsequent lines show matching clones, ranked in order of decreasing significance: number of matches; matching clone (standard format); probability of match being coincidental; shorter distance in bands from end of matching clone to end of its contig; number of bands in incoming clone not found in top matching clone ('d' appended), or number of such difference bands found in subsequent matching clones; canonical status of matching clone or name of its canonical clone, as applicable. If the number of difference bands in the top match exceeds the distance to the end of the contig, a further search is carried out using DIFMAP logic. Usually the output is reviewed with CONTIG9 in AUT mode, and internal clones are rapidly buried. During analysis of this excerpt, F59C6 was found to join contig 545 to contig 755, whilst F59C7 was clearly internal in contig 483 and was buried immediately. Reprinted, with permission, from Sulston *et al.* (18).

program (TAPS) to display the unprocessed image, and then the debugging display program (TARD) to superimpose the standard bands as aligned by the computer. If either the lane following or the alignment is defective (this does not happen if the film is well exposed and free from smudges) the film must be re-scanned or manual corrections must be made [for further details, see (18)]. If the alignment is satisfactory, run the editing program (TAR). Using the cursor, which skips from band to band in each sample lane in turn, delete any unwanted bands (i.e. bands that are too faint, artefactual or outside the area of interest). If there are many unwanted faint bands it is worth raising the adjustable band acceptance threshold for subsequent scans. TAR then adds the normalized band positions to the database.

3.3.2 *Matching*

Write the first and last clones of the new subset in the command file MAP.COM. Submit the file to the batch queue (for subsets of more than a few tens of clones, this job is run overnight). The results are printed out in the form shown in *Figure 4*.

3.3.3 *Analysis*

The objective of analysis is to build contigs of clones that have well established overlaps, to file redundant clones quickly and to record special features, such as genetic positions and problematic areas, in the remark field. Contigs are displayed graphically by means of the analysis program CONTIG9 (*Figure 5*). In the early stages of a project, time should not be wasted on investigation of marginally significant overlaps. Such overlaps can be recorded in the PSS file if the operator so wishes, but many of them will in any case be either confirmed or disproved by the addition of later clones. As the project advances, the AUT mode of CONTIG9 becomes increasingly useful and it is this mode that we shall now describe.

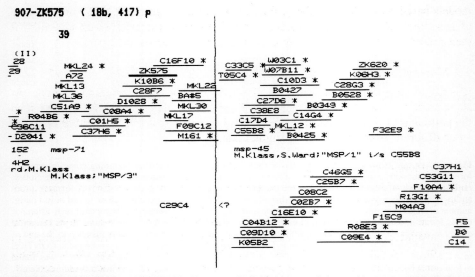

Figure 5. Working display of CONTIG9. The screen is divided into two halves, with current contig displayed above and previous contig below. Top left shows current clone, in standard format (gel number, name, number of bands, contig number). 'p' indicates that this clone has been positioned automatically by the POS routine. The number below the clone name is a band counter for the cursor. Each clone is represented by a line of length proportional to the number of bands in its fingerprint. The line is thickened for the current clone. The asterisk appended to the clone name denotes buried clones. Top left of the contig is a Roman numeral denoting the chromosome, if known. The level immediately below the clones is reserved for gene names (e.g. msp-45); the lower levels contain remarks, which stack down as necessary. The cursor is used for interactive manipulation of the upper contig and clones, and for measurements; it is constrained to move in one band increments. The symbol ' <?' on the lower contig denotes possible join to C29C4. Reprinted, with permission, from Sulston *et al.* (18).

Invoke AUT mode in CONTIG9. As each clone is examined, enter T (for trial mode); the incoming clone will then automatically be displayed in its most likely position in its contig. If the clone is internal, and does not overlie a region that needs additional confirmation (as evidenced by a remark to that effect), then it should usually be 'buried', that is recorded as an associate of the clone that best matches it. If the clone is not internal, verify the exact extent of its overlap with the existing clones by visual inspection of the films; then adjust its position by using the graphics cursor. If no join is apparent from the MAPSUB data, search for one with DIFMAP before proceeding to the next clone.

For future reference, the pattern of overlap between two examined fingerprints can be indicated on the film (using alcohol-soluble ink) by a variety of symbols (e.g. a dot to indicate a positive match, '>' to indicate a band present only in the companion clone, or '0' to indicate a band present only in the marked clone). In this way, when another clone is brought into consideration, the logic of any putative positioning can be deduced.

3.3.4 *Assessment of progress*

A variety of routines have been developed to display the status either of the whole map or of selected subsets of clones. The results can be compared with those generated from models based on random numbers. For example, the program RANCL5 accepts size of genome, size of insert and minimum overlap of two clones to count as a join; it then predicts a map status for any chosen number of clones.

3.4 **Map closure**

It is not to be expected that a map can be completed by purely random methods. At some point it will be necessary to resort to the pre-selection of clones. Depending upon the particular project in hand, the pre-selected clones may usefully be either the same as or different from those in the original bank.

3.4.1 *Different probes and vectors*

In the course of the *C. elegans* project, we have tried a variety of different approaches in an attempt to enhance gap closure.

(i)   The project started with banks made in pJB8. When the Lorist vectors became available, we switched to them in case they might clone in either a more random or a complementary fashion compared with pJB8. In the event a small enhancement was seen but it was clear that more positive methods were required.

(ii)   We then made probes from near the ends of existing contigs, either by using SP6 or T7 promoters in the Lorist vector to yield riboprobes or by restriction digestion to yield DNA fragments. We prepared replica filters carrying Lorist cosmids, probed and picked colonies (see below). Although this approach gave some new joins, it suffered from the drawback that the cosmids that we were trying to find were still under-represented.

(iii) Our current enthusiasm is for cross-probing between Lorist clones from the ends of contigs and YACs, taking advantage of the lack of hybridization between the two vectors. Since YACs are likely to be more random than cosmids, or at the least to have a different distribution, we are hopeful that they will be the most efficient way of linking contigs together.

(iv)   If necessary we can resort to lambda clones as a means of closure. Since lambda grows in a lytic mode we expect initiation of such clones to be less sensitive than that of cosmids to mildly deleterious insert sequences. We know also, in practice, from the work of Andrew Spence (personal communication) that at least one 50 kb gap in the cosmid map can be bridged by walking with lambda clones. However, the time required to carry out such closure on a large number of gaps would be prohibitive, so this approach must be used selectively.

3.4.2 *Screening*

The following protocols apply to both randomly spread and gridded clones. Randomly spreading 5000 cosmid clones on a 14 cm filter allows for accurate picking of single colonies, by means of the device described later. 1000 cosmid or YAC clones can be gridded on a 10 × 8 cm rectangle, selected colonies being tooth-picked into position above a printed grid stuck to the bottom of the plate.

(i) *Preparation of filters.* At present, we use Hybond-N (Amersham) nylon and BA85 (S & S) nitrocellulose filters. Prior to beginning a round of lifting and printing, autoclave

BA85 filters interleaved with squares of 3MM paper, additional squares of 3 MM, and velvet circles slightly smaller than the filters.

(a)   Having spread the required number of colonies, or gridded selected clones, on appropriate agar plates, grow to a suitable size (for Lorist cosmid colonies, we find this to be ~ 13 h at 37°C; for yeast colonies, 12 h at 30°C), then transfer to membranes as follows. (We use BA85 for bacterial colonies, Hybond-N for yeast.)

(b)   Label the filter with a ball-point pen. Lower the filter onto the plate from a marked midline, holding it in a steep 'U' shape for the first contact. Leave the filter in contact with the colonies for a few seconds, peel off and lower onto a fresh plate (colonies up). Make sure that these plates are not too wet by previously leaving them at room temperature for 2 days before storage at 4°C. Incubate for 2 h.

(c)   Nylon 'slaves' are made from these 'masters' as follows.

    (1)   Label a filter (ball-point) and lay it face down to wet on its future growth plate. Perform the master to slave transfer by preparing the following stack (bottom upwards); glass plate, dry 3MM square (can be used several times), master filter (face up), slave filter (face down, start contact with master at midline in same way as for master lift), velvet, and second glass plate. Apply brief hand pressure (20 lb or so), then peel off items in turn, returning both filters to their respective plates, colonies up.

    (2)   When preparing a number of slaves from each of several masters, make up to five successive copies of any given master before returning it to its plate to recover, when the colonies will be seen to re-hydrate within about 20 min. In this way we have been able to make up to 100 copies of master filters of both cosmid and YAC clones with no loss of quality.

    (3)   The master is finally transferred to a glycerol plate (for cosmids: 15 g of agar, 8 g of NaCl, 750 ml of water, 250 ml of glycerol, and appropriate antibiotic; for YACs: 2% agar, 20% glycerol).

    (4)   Place at 4°C for several hours, peel off the filter, place in an empty dish. Store in stacks in plastic bags at −70°C.

(d)   Grow the slaves to give small colonies (10−12 h at 37°C for bacterial colonies, 15−20 h at 30°C for YACs). They can be left at 4°C overnight before processing if convenient.

(e)   Disruption and denaturation of bacterial clones is on saturated 3MM paper sheets (7), either in square plastic boxes for small numbers of filters (25 ml of solution for each 22 × 22 cm 3MM sheet) or directly on a horizontal bench for larger numbers (150 ml of solution for each 46 × 57 cm sheet).

(f)   Lay the filters for 5−10 min on each of the following in turn:

    (1)   10% SDS;

    (2)   0.5 M NaOH, 1.5 M NaCl (change every four cycles);

    (3)   0.5 M Tris−HCl pH 7.4, 1.5 M NaCl;

    (4)   0.5 M Tris−HCl pH 7.4, 1.5 M NaCl;

    (5)   2 × SSPE.

(g)   Dry at room temperature for several hours, and expose to UV light to cross-link the DNA to the filter. Exposure for 5 min to a Biogard safety hood light is recommended.

Yeast clones are treated similarly (with growth incubations at 30°C) except for the following preliminary treatment. Lay the filters for a few minutes on 22 × 22 cm 3MM sheets soaked in 25 ml of 0.8% dithiothreitol in SOE (sorbitol 36 g, 8 ml of 0.5 M EDTA pH 9.0, 2 ml of 1 M Tris−HCl pH 8, add water to 200 ml). Remove the agar from the dishes on which the filters were incubated, wipe and add 5 ml of 0.5 mg/ml zymolyase 20T (Seikagaku Kogyo), and 1% β-mercaptoethanol in SOE. Add a 13 cm 3MM circle, allow it to become wetted, then place the filter on it (avoiding trapped air bubbles) and incubate in sealed plastic bags at 37°C overnight.

(ii) *Probing*. We routinely prepare radiolabelled probes, from whole cosmid or lambda clones or purified fragments, by the 'oligolabelling' procedure of Feinberg and Vogelstein (19,20).

(a)   Leave enough background hybridization — by low stringency washing — for the overall pattern of colonies to be seen. When bringing a new set of filters into use, deliberately over-expose the first autoradiographs so that they can be used subsequently as templates.

(b)   Dry the filters completely (air dry for >2 h) and autoradiograph. To avoid damaging the filters, tape them only lightly or slip them into wallets made by taping tissue paper squares onto a sheet of 3MM (technique devised by Iva Greenwald).

(c)   If necessary remove the probe from the used filters by washing very gently for 10 min in each of: 0.2 M NaOH, 0.5 M NaCl; 0.5 M Tris−HCl pH 7.4, 0.5 M NaCl; 1 × SSPE.

(iii) *Picking colonies from random spreads*. In preparation for picking, align the exposed film with the template and mark each colony with a fine-tipped organic solvent pen. Because the resolution of the camera lucida system is high, it is worth marking the position very precisely — a small dot surrounded by a large circle is best.

(a)   Remove the set of Petri dishes, with the glycerinated master filters, from −70°C to −20°C.

(b)   Set up the camera lucida (*Figure 6*) with dry ice just below the level of the platform in the ice bucket.

(c)   Place the Petri dish on the platform before removing its lid, to prevent condensation on the filter.

(d)   Choosing an easily recognizable edge or sparsely populated area, align the filter with the film; adjust the spotlight so that the colonies appear as glistening points of light superimposed upon the image of the film. Once the film and filter are in register, it is easy to traverse them simultaneously, without losing track of the alignment, to pick each marked colony in turn.

(e)   For picking, moisten a sterile toothpick in an agar plate, touch the colony firmly and scrape it several times in a limited area of the agar plate.

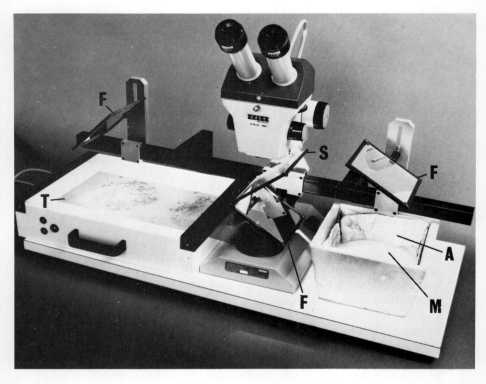

Figure 6. Camera lucida for picking colonies. Fully reflecting mirrors (**F**) are front silvered; beam-splitter mirror (**S**) is semi-silvered on the front and bloomed on the back. For viewing, a magnification of about ×2 at a distance of 50 cm is appropriate; we use either a home-made monocular or a Zeiss SR dissecting microscope fitted with a 500 mm front lens (illustrated). In the former configuration, the filter is illuminated by a small spotlight placed beside the right hand mirror; in the latter the Zeiss spotlight is angled down so as to illuminate the filter via the viewing light path. All the mirrors can be translated and tilted, so that the optical paths to template film (**T**) and master filter (**M**) can be made identical. In use, the ice bucket is filled with dry ice to a level just below a plastic platform that supports the Petri dish containing the master filter. The cold foil (**A**) protects the filter from atmospheric moisture.

(f)   With a fresh stick, cross scrape the first set of streaks in order to isolate single cells. It is convenient to pick about 10 colonies onto one 80 mm plate.

4. CONCLUSIONS

4.1 Status of the *C.elegans* map

At the time of writing, some 90% of the genome has been cloned into cosmids. The number of contigs, having passed through a maximum of 940, fell to about 700 through cosmid linking alone, and has now dropped to 389 by YAC linking. One third of the apparent gaps between cosmid contigs proved not to be gaps at all but small overlaps that had not yet been detected; but a large number of real gaps exist for which cosmids are rare or unobtainable. At this point, a number of researchers are being helped by the larger contigs (there are more than 95 that are >200 kb, totalling ~60% of the genome, and 9 that are >1000 kb). Others, however, are frustrated by finding themselves in regions with small contigs and gaps. We are thereby being made acutely aware of the importance of proceeding to full linkage as fast as possible.

Thus, our experience suggests that cosmids are not without drawbacks for eukaryote mapping, although they have proved admirable for at least one prokaryote (see below). We anticipate that YACs will provide the fastest route to full linkage of the map, and that lambda clones can subsequently be picked up as required for more precise cloning of gaps in the cosmid map.

4.2 **Other projects**

Although an exhaustive review would be inappropriate in this book, a short list of other mapping projects will serve to put the detailed procedures described above into perspective.

4.2.1 *Escherichia coli*

4×10^6 bp. Y.Kohara *et al.* (21). Lambda clones; partial digestion by eight different enzymes; analysis on agarose, with indirect end labelling. True restriction map for entire genome.

4.2.2 *Rhodobacter capsulatus*

3×10^6 bp. J.Williams and S.Brenner (personal communication). Cosmid clones; fingerprinting technique, as for *C.elegans*, but using *Bam*HI instead of *Hind*III to allow for the high GC content. Map complete.

4.2.3 *Saccharomyces cerevisiae*

12×10^6 bp. M.Olson and co-workers (22). Lambda clones; simultaneous digestion by *Hind*III and *Eco*RI, analysis on agarose. Close to 100% of genome cloned, a few hundred contigs (Olson, personal communication).

4.2.4 *Arabidopsis thaliana*

70×10^6 bp. B.Hauge, H.Goodman *et al.* (personal communication). Cosmid clones; fingerprinting technique, as for *C.elegans*.

5. GENERAL RECIPES

(i) TE
10 mM Tris−HCl pH 7.4, 0.1 mM EDTA.

(ii) 10 × ligation buffer
0.5 M Tris−HCl, pH 7.4, 0.1 M MgCl$_2$.

(iii) 10 × medium salt restriction buffer
500 mM NaCl, 100 mM Tris−HCl pH 7.4, 100 mM MgCl$_2$, 10 mM dithiothreitol.

(iv) 10 × TBE

Per litre: 108 g of Tris base; 55 g of boric acid; 9.3 g of EDTA.

(v) 50 × TAE

242 g of Tris base; 57.1 ml of glacial acetic acid; 100 ml of 0.5 M EDTA pH 8.0; water to 1 litre.

(vi) 4% denaturing gel solution

480 g of urea; 38 g of acrylamide; 2 g of bis-acrylamide; water to approximately 800 ml. Dissolve with minimal heat. Add 20 g of Amberlite MB-3 resin; stir gently for 1 h. Filter using a No. 2 glass sinter funnel. Add 100 ml of 10 × TBE and water to 1 litre.

(vii) Formamide dyes (denaturing dye mix)

100 ml of de-ionized formamide; 0.1 g of xylene cyanol FF; 0.1 g of bromophenol blue; 0.3 g of EDTA.

(viii) Boiled RNase

100 mg of RNase A; 10 ml of 10 mM Tris−HCl pH 7.4, 15 mM NaCl. Incubate at 100°C for 15 min. Cool, aliquot and store at −20°C.

(ix) 2 × TY medium

Per litre: 16 g of tryptone; 10 g of yeast extract; 5 g of NaCl. Adjust to pH 7.4 (~2 ml of 4 M NaOH).

(x) CY medium

Per litre: 10 g of casamino acids; 5 g of yeast extract; 3 g of NaCl; 2 g of KCl. Adjust to pH 7.0.

(xi) 20 × SSPE

104 g of NaCl; 15.6 g of $NaH_2PO_4.2H_2O$; 3.7 g of $Na_2EDTA.2H_2O$; 25 ml of 4 M NaOH and water to 500 ml.

(xii) Lambda dil

10 ml of 1 M Tris−HCl pH 7.4; 5 ml of 1 M $MgSO_4$; 11.7 g NaCl; 1 g gelatin and water to 1 litre.

6. ACKNOWLEDGEMENTS

Explicit contributions to the text from Robert Waterston, Toby Gibson, George Brownlee and Vroni Knott have already been noted. Robert Waterston and, more recently, Yuji Kohora are active collaborators on the *C.elegans* project. We thank Humaira Ameer, Jane Kiff and Caterina Randolf for technical assistance. In addition, we have received a great deal of encouragement, practical support and useful comment from numerous colleagues. We would like to thank particularly Donna Albertson, Sydney Brenner, Iva Greenwald, Jonathan Karn, Peter Little, Maynard Olson and the many *C. elegans* researchers who have freely shared their clones and their data with us.

7. REFERENCES

1. Coulson,A., Sulston,J., Brenner,S. and Karn,J. (1986) *Proc. Natl. Acad. Sci. USA*, **83**, 7821.
2. Collins,J. and Hohn,B. (1978) *Proc. Natl. Acad. Sci. USA*, **75**, 4242.
3. Ish-Horowicz,D. and Burke,J.F. (1981) *Nucleic Acids Res.*, **9**, 2989.
4. Gibson,T., Coulson,A., Sulston,J. and Little,P.F.R. (1987) *Gene*, **53**, 275.
5. Gibson,T., Rosenthal,A. and Waterston,R. (1987) *Gene*, **53**, 283.
6. Little,P.F.R. and Cross,S.H. (1985) *Proc. Natl. Acad. Sci. USA*, **82**, 3159.
7. Maniatis,T., Fritsch,E.F. and Sambrook,J. (1982) *Molecular Cloning: A Laboratory Manual*. Cold Spring Harbor Laboratory Press, Cold Spring Harbor, New York.
8. Karn,J., Matthes,H.W.D., Gait,M.J. and Brenner,S. (1984) *Gene*, **32**, 217.
9. Burke,D.T., Carle,F.G. and Olson,M.V. (1987) *Science*, **236**, 806.
10. Sulston,J.E. and Brenner,S. (1973) *Genetics*, **77**, 95.
11. Sternberg,N., Tiemeier,D. and Enquist,L. (1977) *Gene*, **1**, 255.
12. Rosenberg,S.M. (1985) *Gene*, **39**, 313.
13. Cami,B. and Kourilsky,P. (1978) *Nucleic Acids Res.*, **5**, 2381.
14. Carle,G.F. and Olson,M.V. (1984) *Nucleic Acids Res.*, **12**, 5647.
15. Birnboim,H.C. and Doly,J. (1979) *Nucleic Acids Res.*, **7**, 1513.
16. Gibson,T.J. and Sulston,J.E. (1987) *Gene Anal. Technol.*, **4**, 41.
17. Garrof,H. and Ansorge,W. (1981) *Anal. Biochem.*, **115**, 450.
18. Sulston,J., Mallett,F., Staden,R., Durbin,R., Horsnell,T. and Coulson,A. (1988) *Cabios*, **4**, 125.
19. Feinberg,A.P. and Vogelstein,B. (1983) *Anal. Biochem.*, **132**, 6.
20. Feinberg,A.P. and Vogelstein,B. (1984) *Anal. Biochem.*, **137**, 266.
21. Kohara,Y., Akiyama,K. and Isono,K. (1987) *Cell*, **50**, 495.
22. Olson,M.V., Dutchik,J.E., Graham,M.Y., Brodeur,G.M., Helms,C., Frank,M., MacCollin,M., Scheinman,R. and Frank,T. (1986) *Proc. Natl. Acad. Sci. USA*, **83**, 7826.
23. Olson,M.V., Loughney,K. and Hall,B.D. (1979) *J. Mol. Biol.*, **132**, 387.
24. Knott,V., Rees,D.J.G., Cheng,Z. and Brownlee,G.G. (1988) *Nucleic Acids Res.*, **16**, 2601.

CHAPTER 3

Pulsed-field gel electrophoresis and the technology of large DNA molecules

C.L.SMITH, S.R.KLCO and C.R.CANTOR

1. INTRODUCTION

The advent of pulsed-field gel (PFG) electrophoresis (1) allowed the direct resolution of chromosomal DNAs from a variety of yeasts and parasitic protozoa (2−3), because these molecules range in size from 0.2 to 10×10^6 base pairs (Mb). Subsequent work has resulted in the development of techniques for generating and manipulating DNAs in this size range from a variety of both prokaryotic and eukaryotic organisms (4−7). Thus, it is now possible to do megabase molecular studies on whole chromosomes from virtually any organism. Such experiments have resulted in complete physical maps for large regions of mammalian chromosomes (8) and of the entire *Escherichia coli* chromosome (9). The purpose of this chapter is to review the biochemical methods used for generating large DNA molecules and the PFG methods used for analysing such large DNA molecules. Reviews of related topics may be found elsewhere including PFG techniques (10), PFG theory (11) and applications of PFG (12).

2. PREPARATION OF DNA SAMPLES

Very high molecular weight DNA samples are impossible to prepare by conventional solution methods because they are extremely sensitive to shear damage. The enormous axial ratio of these molecules accounts for this sensitivity. For example, human chromosome 21 is the smallest human chromosome. It is estimated to be 50 Mb in size. The DNA of this chromosome is a single molecule 15 mm in length, yet the diameter of a DNA helix is only 20 nm. Thus, this DNA molecule has an enormous axial ratio of 10^6.

In order to prevent shear damage, procedures have been developed for extracting DNA from cells embedded in agarose. The procedures are extremely simple, and the samples can be used in conventional electrophoretic and cloning experiments. Basically, freshly grown live cells are suspended in liquid low-gelling agarose. The agarose is allowed to solidify in a 100 μl mould to form blocks called inserts (*Figure 1*). The moulds can be purchased from Pharmacia-LKB or conveniently made by cementing together a set of gel electrophoresis sample combs in a head to tail manner. Care should be taken in storage and handling of the mould so that nuclease contamination is avoided. Before use, the mould should be rinsed extensively with distilled water and dried with Kimwipes while wearing gloves. One side of the mould is then covered with tape.

As an alternative to solid agarose inserts, the agarose can be solidified into beads

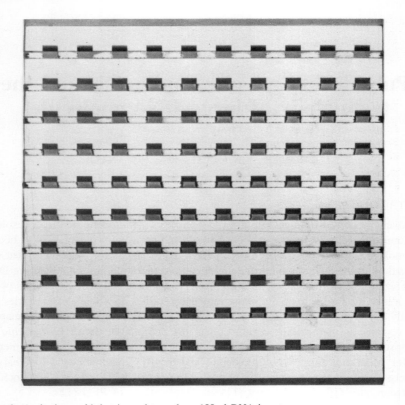

Figure 1. A plastic mould that is used to make a 100 μl DNA insert.

by pipetting the agarose−cell suspension into protol oil while vortexing (13). The size of the beads is controlled by the vortexing speed. DNA samples in beads can be manipulated like ordinary liquid samples. However, we prefer to use insert preparations because of the ease in manipulating them as described below.

To prepare DNA from cells containing no walls, the insert samples are simply put into a solution of ESP (see *Table 1*) at 50°C for 2 days (*Tables 2* and *3*). This solution will lyse the cells, and remove all proteins and other molecules from the DNA. Intact chromosomal DNA samples amenable to analysis by PFG electrophoresis are simply loaded onto gels directly from ESP. Insert DNA samples in ESP may be stored indefinitely in ESP at 4°C or even shipped at room temperature in Eppendorf tubes.

Cell wall material must be removed before samples are put into ESP. Protocols for preparing DNA from organisms with cell walls are shown in *Tables 4−6*. Cell wall material varies somewhat with each specific organism. In some cases it may be necessary to try a variety of enzymes which degrade polysaccharide cell wall material. Many of these enzymes do not have divalent cation requirements. Thus incubation of cells with these enzymes can be carried out in high concentrations of ethylenediamine tetraacetic acid (EDTA) to help protect the chromosomal DNA.

All samples should be finally incubated in ESP for 2 days at 50°C. The inserts should

Table 1. Solutions for sample preparation and PFG electrophoresis.

1. *50 × Denhardt's stock*

    This solution is a component of the pre-hybridization and hybridization mixes.

	per 500 ml
Polyvinylpyrrolidone (Sigma PVP-40)	5 g
Ficoll (Sigma F-9378)	5 g
BSA Pentax Fraction V (Sigma A-4503)	5 g

    Filter through disposable Nalgene filters.
    Store at −20°C in 30 ml aliquots.

2. *EC lysis*

    The abbreviation stands for *E.coli* (EC) lysis. The solution is used for making spheroblasts from a variety of bacteria.

Final concentration	per litre	
6 mM Tris−HCl (pH 7.6)	6 ml	1 M stock
1 M NaCl	58.4 g	
100 mM EDTA (pH 7.5)	100 ml	0.5 M
0.5% Brij-58 (Sigma P5884)	50 ml	10% (or 5 g)
0.2% Deoxycholate (Sigma D6750)	50 ml	4%
0.5% Sarkosyl (Ciba Geigy NL 30)	5 ml	100%

    Autoclave
    Add fresh: 1 mg/ml hen egg white lysozyme
    　　　　　 20 µl/ml bovine pancreatic RNase[a] (2 µl of 10 mg/ml stock per ml of EC lysis)

3. *ESP*

    The abbreviation derives from EDTA (E), sarcosine (S), proteinase K (P). This solution is used for all large DNA preparations. It has also been called NDS.

    Final concentration
    0.5 M EDTA (pH 9−9.5)
    Autoclave
    1% Lauroyl sarcosine (Sigma L-5125)
    Shake vigorously to get into solution
    1 mg/ml proteinase K
    Incubate at 37°C for 2 h
    Freeze in aliquots
    To use: dilute 1/2 with sample and incubate for 1−2 days at 50°C.

4. *Hybridization and pre-hybridization mix*

Final concentration		per litre	
3 ×	SSC	100 ml	30 × stock
0.1%	SDS	5 ml	20%
10 ×	Denhardt's	200 ml	50 ×
10%	Dextran sulphate[b]	200 ml	50%
	Herring sperm DNA[c]		5 mg/ml

    Store frozen in 30 ml aliquots for pre-hybridization (one aliquot per 20 cm × 20 cm filter), and in 20 ml aliquots for hybridization (one aliquot per 20 cm × 20 cm filter).

5. *10 × M 9*

    per litre
    50 g Na₂HPO₄
    30 g KH₂PO₄
    　5 g NaCl
    10 g NH₄Cl

Dilute to 1 × and autoclave.
After autoclaving add:

Final concentration	per litre	
1 mM $MgSO_4$	1 ml	1 M stock
0.1 M $CaCl_2$	0.2 ml	0.5 M
0.4% Glucose	10 ml	40%
0.5% Casamino acids[d]	25 ml	20%
Optional additions:		
50 μg/ml Amino acids	10 ml	5 mg/ml
10 μg/ml Vitamins	10 ml	1 mg/ml
50 μg/ml Bases	25 ml	2 mg/ml

6. *Pett IV*

This solution is used to wash bacterial cells prior to spheroblasting.

Final concentration	per litre	
10 mM Tris−HCl (pH 7.6)	10 ml	1 M stock
1 M NaCl	58.44 g	

Autoclave

7. *0.1 M PMSF*

This is to inactivate Proteinase K.
Re-suspend 17.5 mg of PMSF in 1 ml of isopropanol.
Note that PMSF is unstable, and this solution must be made fresh each day. PMSF is also very toxic.

8. *PSG*

The abbreviation derives from phosphate (P), saline (S), glucose (G). This solution is used for washing protozoa prior to lysis.

Final concentration	per litre	
75 mM $NaPO_4$ (pH 8)	75 ml	1 M stock
65 mM NaCl	13 ml	5 M
10% Glucose	100 g	

Autoclave

9. *RBC lysis buffer*

This solution is used in the preparation of large DNA from fresh whole blood.

	per litre
1 mM NH_4HCO_3	79 mg
114 mM NH_4Cl	7.6 g

Autoclave

10. *SSC*

The abbreviation stands for standard saline citrate. This solution is used for blotting and hybridization.

20 × SSC	per litre
3.0 M NaCl	1051.8 g
0.30 M Na_3citrate.$2H_2O$	529.2 g

Adjust to pH 7.0 with concentrated HCl.

11. *TE*

The abbreviation derives from Tris (T) and EDTA (E). This solution is used for DNA preparations.

Final concentration	per litre	
10 mM Tris−HCl (pH 7.4)	10 ml	1 M stock
0.1 mM EDTA	0.2 ml	0.5 M

12. *TBE*

The abbreviation derives from Tris (T), borate (B), EDTA (E). This solution is used as the buffer in PFG electrophoresis. Note that the EDTA concentration is 10-fold lower than conventional formulations.

Final 1 × concentration	per 6 litres for 10 × stock
0.1 M Tris base	726.6 g
0.1 M Boric acid	370.8 g
0.2 mM Na$_2$EDTA.2H$_2$O	
(mol. wt 372.2)	4.47 g

13. *YPD*

The abbreviation derives from yeast (Y), peptone (P) and dextrose (D). This solution is the culture medium for *S.cerevisiae* and *S.pombe*.

Final concentration	per litre
1% Bacto-yeast extract	10 g
2% Bacto-peptone	20 g
2% Dextrose	20 g
2% Bacto-agar	20 g

14. *Zymo-buffer*

This solution is used in the preparation of yeast spheroblasts.

Final concentration	per 100 ml	
20 mM Citrate-phosphate buffer (pH 5.6)	10 ml	0.2 M stock
50 mM EDTA (pH 5.6)	10 ml	0.5 M
0.9 M Sorbitol	164 g	

0.2 M Citrate-phosphate buffer

Mix together
11.6 ml 0.4 M Na$_2$HPO$_4$
8.4 ml 0.2 M citric acid

15. *2 mg/ml Zymolyase 100 T (Calbiochem)*

This solution is used for preparing yeast spheroblasts.
Make up in sterile 10 mM NaPO$_4$ (pH 7.5) + 50% glycerol
Store at −20°C.

---

[a]Incubate the RNase stock solution for 20 min at 80°C before storing frozen at −20°C.
[b]Sigma D-0768; 5000 average molecular weight.
[c]Shear DNA by sonication to 1 × 10^6 bp. Boil for 10 min, chill quickly in an ice water bath, then add to mix. (Add 20 ml of stock solution per litre to hybridization mix, and 50 ml of stock solution per litre to pre-hybridization mix, to give final concentrations of 100 and 250 mg/ml, respectively).
[d]One may autoclave 1 × M9 with casamino acids instead of adding them later. Alternatively, one may add individual amino acids separately instead of using casamino acids (see optional additions). Casamino acids are low in aromatic amino acids, such as tryptophan. This amino acid should be added if required by the particular strain.

Table 2. Preparation of DNA insert samples from protozoa.

1.	Suspend the cells in PSG.
2.	Dilute with an equal volume of liquified 1% low gelling temperature agarose.
3.	Mix gently.
4.	Distribute with a pipetteman or Pasteur pipette into the mould covered on one side with tape. Avoid air bubbles.
5.	Put the mould with the cells in a freezer ($-20°C$) for exactly 5 min (otherwise samples freeze) or $10-15$ min at $4°C$.
6.	Remove the tape.
7.	Push the inserts out with a sterile bent Pasteur pipette into at least twice the volume of ESP. (The inserts should push out easily and be nicely formed. If not, cool for another few minutes.)
8.	Incubate at $50°C$ in ESP for 2 days with gentle shaking. Samples should clear indicating lysis.
9.	Store at $4°C$ in ESP (may be shipped in ESP at room temperature).

Organism	Cells/insert (100 µl)	Load/lane
Trypanosoma brucei	2×10^8	1/16
Plasmodium falciparum	5×10^7	1/8
Giardia lambia	1×10^7	1/10
Leishmania	2×10^8	1/16

clear earlier than this, indicating lysis of the cells. However, the 2-day incubation appears to be necessary to remove bound material from the DNA molecules completely. DNA molecules with bound protein may have altered electrophoretic mobility. Additionally, samples which have been incubated for shorter time periods have been resistant to some restriction enzyme digestions.

The most common problem encountered in applying the preparation techniques described below to a new organism is controlling the final DNA concentration. The goal is a DNA concentration which allows a convenient fraction of an insert to be loaded onto a gel. The concentration needed to achieve this goal must be determined empirically. Suggestions for cell numbers and DNA amounts are given in the protocols. These may need to be adjusted for culture conditions, and one should not be afraid to adjust them. For preliminary experiments with new organisms, usually we recommend making samples at three concentrations, 5-fold less, the same and 5-fold more than those shown in *Tables 2−6*.

Chromosomal DNA molecules are exquisitely sensitive to nucleases because of their large target size. A good rule of thumb is to treat all solutions and materials that come into contact with the samples as if one were working with RNA. Solutions should be aliquoted and sterilized in small quantities. All equipment that comes in contact with the sample should be sterilized. The samples may be pushed from the insert mould or from one tube to another with an alcohol-flame-sterilized bent glass rod. The sample may also be handled with a glass coverslip while wearing gloves. No DNA sample should ever be handled with anything containing metal such as stainless steel spatulas or razor blades. DNA solubilizes and binds to many divalent metals very strongly and exposure to these metals usually results in strand breakage.

2.1 Preparation of DNA insert samples from protozoa (*Table 2*)

Free living unicellular protozoa may be used directly to make insert DNA preparations.

Table 3. Preparation of DNA insert samples from mammalian cells.

A. *Cells in culture*
1. Harvest almost confluent or early stationary phase cells.
2. Pellet the cells at 800 r.p.m. in a Beckman TJ6 centrifuge for 10 min at room temperature and wash twice with PBS.
3. Resuspend the cells at 2×10^7 cells/ml (2×10^6 cells/insert).
4. Mix with an equal volume of 1% low gelling agarose.
5. Proceed as in *Table 2*, steps 4−9.

DNA concentration:
We assume the final concentration is approximately 10 µg/insert (or 100 µg/ml).
Usually 1/6 of an insert is loaded per gel lane.

B. *DNA insert preparation from fresh blood*
1. Collect whole blood by filling several 10 ml lavender top Vacutainer tubes (BD 6457) containing 15 mg K_3EDTA.
2. Transfer approximately 13 ml of blood to a 50 ml conical centrifuge tube.
3. Add RBC lysis buffer to the 50 ml mark and mix by inverting the tube.
4. Leave on ice for 30 min.
5. Spin for 15 min at 3000 r.p.m. at 4°C in a Beckman TJ6 centrifuge.
6. Decant the supernatant by carefully pouring it off.
7. Re-suspend the pellet in 25 ml of RBC lysis buffer with gentle vortexing.
8. Leave on ice for 5 or 10 min to complete lysis.
9. Spin for 15 min at 3000 r.p.m. at 4°C in a Beckman TJ6 centrifuge.
10. Re-suspend the pellet in 20 ml of RBC lysis buffer with gentle vortexing and incubate on ice for 5 min.
11. Spin again at 3000 r.p.m. at 4°C in a Beckman TJ6 centrifuge.
12. Re-suspend the pellet in 20 ml of cold PBS.
13. Spin for 10 min at 2500 r.p.m. at 4°C in a Beckman TJ6 centrifuge.
14. Decant the supernatant and re-suspend the pellet containing the white cells in 1.5 ml of PBS. (Some red pigments may remain with the white cells.)
15. Add 1.7 ml of 1% low gelling temperature agarose, mix gently and distribute into moulds as described in *Table 2*, step 4.
16. Continue as for steps 5−9, *Table 2*.
17. DNA concentration: a normal white blood cell count is approximately 7000 cells/mm^3 or 7×10^6 cells/ml. The above procedure results in about a 10% loss of white blood cells. Thus, the final number of cells per insert (100 µl) is about 1×10^6 or approximately 10 µg DNA (100 µg/ml).
18. Load 1/8 per gel lane.

C. *DNA preparation from lymphocytes prepared from fresh blood using Ficoll*
1. Dilute blood (10 ml) with 3 vols of PBS in a 50 ml conical centrifuge tube.
2. Layer 10 or more millilitres of Histopaque (Sigma 1077) under the diluted blood.
3. Spin at 3000 r.p.m. for 15 min. (The saline remains above the Ficoll; the cloudy interface is composed of lymphocytes.)
4. Aspirate off all but 10 ml of the saline.
5. Remove the remaining saline and lymphocytes with a pipette aspirating in a circular pathway.
6. Count the cells. The yield is usually around 75%.
7. Make inserts at 1×10^6 cells/insert as in A above.

They can be pelleted out of the culture medium by centrifugation. Intracellular parasites should be purified away from blood cells or host cells before inserts are made. Erythrocytes may be lysed by the protocol given in *Table 3B*. In some cases trypanosomes also have been purified away from various blood elements on DEAE columns (14).

Table 4. Preparation of *E.coli* DNA inserts.

Growth and harvesting

1. Grow *E.coli* to 1.5×10^8 cells/ml (Klett = 45 using a red 66 filter) in 10 ml of M9 plus appropriate supplements. Aerate well at 37°C.
2. Add 18 μl of 0.1 g/ml chloramphenicol (made up in 95% ethanol) to a final concentration of 180 μg/ml.
3. Continue incubating and monitoring growth for another hour.
4. Chill by swirling in an ice bath.
5. Spin at 8000 r.p.m. for 15 min at 4°C in a Sorvall SS34 rotor.
6. Resuspend in 10 ml of Pett IV.
7. Spin at 8000 r.p.m. for 10 min at 4°C in a Sorvall SS34 rotor.

Insert making (30 insert samples at 0.5 μg DNA/insert)

1. Re-suspend the cells thoroughly in 1.6 ml of Pett IV.
2. Warm the cells to 30−40°C.
3. Dilute with an equal volume of 1% low gelling temperature agarose made up in sterile water.
4. Distribute into the mould, covered on one side with tape. Avoid air bubbles.
5. Cool the mould at −20°C for 5 min exactly or at 4°C for 10−15 min. Then push into an equal volume of EC lysis solution. The inserts should push out easily and be nicely formed. If not, cool for another few minutes.
6. Incubate overnight at 37°C with gentle shaking.
7. Discard the solution and incubate the insert in an equal volume of ESP for 2 days at 50°C with gentle shaking.
8. Store the inserts at 4°C or ship at room temperature in ESP.
9. Usually 1/6 of an insert is loaded per gel lane.

To calculate yield (μg) of DNA from actual Klett reading and volume (vol):

$$(\text{Klett}) (\text{vol}) \left(\frac{10^8 \text{ cells/ml}}{30 \text{ Klett units}} \right) \left(\frac{2.5 \text{ chromosomes}}{\text{cell}} \right) \left(\frac{1 \ \mu\text{g}}{2.4 \times 10^8 \text{ chromosomes}} \right) =$$

or $(\text{Klett}) (\text{vol}) (0.0362) = \text{total } \mu\text{g of DNA}$

For example, a 10 ml Klett 45 *E.coli* K12 culture has $45 \times 10 \times 0.0362 = 16 \ \mu$g of DNA.

2.2 Preparation of DNA from cultured cells (*Table 3A*) and white blood cells (*Table 3B,C*)

The procedure for preparing DNA from cells in culture is very similar to that described for protozoa, since these cells do not contain walls. For restriction mapping experiments one should be careful to use cell lines that have a normal genome. For instance, HeLa cells are easily grown but contain numerous chromosomal rearrangements and variable chromosome ploidies. One million diploid cells should contain 6.6 μg of DNA, assuming a normal genome size of 6×10^9 bp of DNA. (Since unsynchronized cultured cells are a mixture of diploid, tetraploid and intermediate states, we usually treat 1×10^6 cells as roughly containing 10 μg of DNA.)

The procedure in *Table 3A* has also been used to obtain intact chromosomal samples from various insect cell lines. The cell concentration should be adjusted accordingly as the genome size is ordinarily 10- to 20-fold less than a mammalian genome.

The procedure used for white blood cells is essentially the same as that for cultured cells once they are free of contaminating erythrocytes. Erythrocytes may either be lysed (*Table 3B*) or removed by centrifugation of whole blood through Ficoll or Percoll gradients. A procedure we have used is shown in *Table 3C*.

Table 5. Preparation of *Saccharomyces cerevisiae* DNA inserts.

1.	Grow the cells overnight at 30°C in 200 ml of YPD.
2.	Spin at 6000 r.p.m. in a Sorvall GSA rotor, or equivalent.
3.	Re-suspend in 15 ml of 50 mM EDTA (pH 7.5). Determine the cell concentration by reading the optical density of a 10^{-3} dilution using a Klett 66 red filter[a].
4.	Spin at 2000 r.p.m. for 10 min in a Beckman TJ6 rotor or equivalent in 50 ml conical tubes.
5.	Re-suspend the cells at 1×10^{10} cells/ml in 50 mM EDTA (pH 7.5).
6.	Insert making

At 38°C mix together:

2 ml	1×10^{10} cells/ml
0.04 ml	2 mg/ml Zymolyase 100T
2 ml	1% low gelling temperature agarose.

Keep at 38°C while distributing into 100 μl moulds.

7.	Make spheroplasts by pushing inserts out of the mould into an equal volume of 0.5 M EDTA (pH 7.5) containing 7.5% β-mercaptoethanol and incubating overnight at 37°C with gentle shaking (Yes, 7.5 ml of 14 M β-mercaptoethanol per 100 ml).
8.	Remove the solution and rinse the inserts with 50 mm EDTA (pH 9−9.5).
9.	Lyse the cells by transferring the spheroplast to a solution of ESP and incubating for 2 days at 50°C with gentle shaking.
10.	Store in ESP at 4°C.
11.	Usually 1/5 is loaded per gel lane.

[a]To calculate cell concentration: Klett 50 = 5×10^7 total cells.
Klett $\times 10^3$ (dilution factor) $\times 5 \times 10^7$ cells/ml = total cells /ml.

2.3 Preparation of DNA from prokaryotes (*Table 4*)

A sample protocol for preparing DNA from *E.coli* is shown in *Table 4*. This protocol has been used successfully to obtain intact chromosomal DNA amenable to digestion with essentially all restriction enzymes from a wide assortment of bacteria and Archaebacteria, including *Salmonella, Legionella, Mycobacterium, Haemophilus, Bacillus, Streptomyces, Halobacterium* and a few unknown fast-growing contaminants. Chromosome sizes in bacteria vary from 1 Mb to 20 Mb and chromosome number per cell can vary from one to five during different culture conditions. Thus, preliminary experiments should involve testing a wide range of cell concentrations.

In some cases it may be necessary to make minor adjustments in the protocol. For instance, the concentration of NaCl was raised to 0.5 M in Pett IV when this procedure was used with *Halobacterium* in order to prevent premature lysis of the *Halobacterium* cells.

The procedure in *Table 4* is designed to synchronize replication forks by incubation of cells in chloramphenicol for 1 h before harvesting. This allows ongoing rounds of replication to finish, but does not permit new rounds of replication to initiate. It is also possible to prepare DNA samples from unsynchronized overnight cultures. However, one should bear in mind that regions of the chromosome near the replication terminus may be under-represented in such preparations (15). For many experiments, this does not matter. However, ethidium bromide staining intensity is mass dependent. Thus, chromosomal DNA near the replication terminus region may be difficult to see by ethidium bromide.

Table 6. Preparation of *Schizosaccharomyces pombe* DNA inserts.

1.	Streak the yeast strain on a YPD plate; incubate overnight at 30°C.
2.	Inoculate a single colony into 300 ml of YPD medium. Incubate with shaking at 30°C for about 25 h (until the culture density is ~450 Klett units using a red 66 filter).
3.	Cool the cells on ice for 10 min.
4.	Spin at 2000 r.p.m. for 10 min in a Beckman TJ6 centrifuge.
5.	Re-suspend the pellet in 75 ml of 50 mM EDTA (pH 7.5).
6.	Spin at 3000 r.p.m. for 10 min in a Beckman TJ6 centrifuge.
7.	Re-suspend the cells in 25 ml of Zymo-buffer, and add 3.8 ml of 2 mg/ml Zymolyase 100T (final concentration is 0.3 mg/ml).
8.	Incubate at 37°C for 2−3 h. Check the formation of the spheroplasts by adding an equal volume of 1% sodium dodecyl sulphate (SDS) and observing for cell lysis with a microscope. Count the number of cells under the microscope.
9.	Spin at 5000 r.p.m. for 10 min in a Beckman TJ6 centrifuge.
10.	Re-suspend the spheroplasts in Zymo-buffer to a final concentration of 2×10^9 spheroplasts/ml. (Usually ~6.5 ml of buffer is added.)
11.	Add an equal volume of 1% low melting temperature agarose and mix well.
12.	Distribute into moulds and allow to solidify.
13.	Push the inserts into ESP and incubate for 1 day at 50°C with gentle shaking.
14.	Change ESP to a fresh solution and incubate for another day at 50°C before storing at 4°C.
15.	Usually 1/8 is loaded per gel lane.

2.4 **Preparation of DNA from fungi** (*Tables 5 and 6*)

Two protocols are given for yeast. The first protocol in *Table 5* is for *Saccharomyces cerevisiae*, while the second is for the fission yeast *Schizosaccharomyces pombe*. Both use Zymolyase 100T to remove the cell wall and make spheroplasts. It is more difficult to make spheroblasts with *S.pombe*. Thus we recommend their production be monitored microscopically. Spheroblasts are sensitive to hypotonic conditions and small amounts of detergent (0.5%). One can monitor this sensitivity by watching cells lyse under the microscope upon addition of detergent.

Other fungi may have different cell wall components. There are a variety of Zymolyases which differ in their activity and specificity. Thus it may be useful to try different ones. We have observed that there is too much nuclease in glucylase for this enzyme to be useful in the preparation of large DNA. Novozyme, produced by Novo Bio Labs, is reported to be a mixture of different enzymes which act on fungal cell walls. We have not tried it because we were somewhat discouraged by the fact that it was brown in colour. A new, more highly purified version may soon be available.

2.5 **Preparation of DNA from multicellular organisms, algae and plants**

Experiments are under way to work out procedures for obtaining intact chromosomal DNAs from solid tissues. Preliminary experiments with filarial worms, *Caenorhabditis elegans, Drosophila* embryos, and *Chlamydomonas* suggest that a variety of procedures may be used. Small multicellular organisms may simply be treated as described for unicellular organisms (*Tables 2* and *3*). Treatment of mobile cells with 0.2 mM sodium azide (or KCN) may be necessary to prevent them from leaving the inserts. Preliminary experiments using dounced, collagenase-treated, trypsinized or simply cut up material

Table 7. Restriction enzyme digestion of DNA insert preparations.

1.	Wash 10 inserts in each solution of the following solutions in a 15 ml conical disposable tube for 2 h to overnight on a rotator at low speed or with gentle shaking at room temperature.
	(i) Twice with 10 ml of TE containing 1 mM PMSF (add fresh 100 μl of 0.1 M PMSF to 10 ml of TE).
	(ii) Three times with 10 ml of TE (no PMSF).
2.	Set up digestion in an Eppendorf tube in a final reaction volume twice that of the insert volume.

1 insert	100 μl
10 × assay buffer[a]	20 μl
20 mg/ml bovine serum albumin	1 μl
Water plus enzyme[b]	79 μl
Total	200 μl

3.	Aspirate off the buffer.
4.	Add 1 ml of ES (ES is ESP without proteinase K).
5.	Incubate for 2 h to overnight at 50°C.
6.	Aspirate off the ES and add 250 μl of ESP and continue incubating at 50°C for 2 h to overnight.
7.	Store at 4°C in ESP.

[a]Add fresh sulphydryl reagent; keep glycerol at less than 5%.
[b]Add 20 enzyme units per μg DNA.
Incubate overnight at 37°C or appropriate temperature with gentle shaking.

placed in ESP have all yielded high molecular weight DNA. One can also first isolate nuclei (16) and then prepare insert samples from them. More experiments are needed to determine which of the methods is best for particular samples.

3. MANIPULATION OF DNA INSERT SAMPLES

For many applications it is easier to manipulate DNA in agarose than in solution. High molecular weight DNA in agarose is certainly easier to handle than viscous solution preparations. Small molecular weight components (salts, proteins and even 5-kb DNA molecules) simply diffuse freely in or out of the insert samples with gentle agitation. We treat the agarose insert samples simply as a 100 μl volume sample.

The ability to manipulate DNA enzymatically in agarose depends on the batch of agarose. At least half of the batches made contain impurities which prevent enzymes from working efficiently. FMC Corporation sells low gelling temperature agarose (INCERT) specifically tested for suitability for restriction enzyme digestion.

3.1 Enzymatic manipulation

In order to manipulate DNA insert samples with enzymes it is necessary to inactivate any proteinase K remaining from the sample preparation and to dialyse away the EDTA and detergent (*Table 7*). Proteinase K is an extremely hardy enzyme and will remain active for at least 1 year in ESP at 4°C. However, it is completely inactivated by treatment with phenylmethylsulphonyl fluoride (PMSF), which is extremely unstable and must be made fresh each day. Any remaining proteinase K left in inserts will destroy added enzymes. One should also be careful not to contaminate the laboratory bench, pipetteman and other items with proteinase K.

Most if not all restriction enzymes, from a variety of sources, work well in the right batch of agarose. When enzymes do not work well in agarose, they usually will not work well in solution experiments. Routinely we use an enzyme to DNA ratio of 20:1 (Units:μg). In many cases this may be a 10-fold excess over what is really needed. However, in our experience, with a wide variety of DNA samples, enzymes and experimenters, this protocol almost always guarantees complete digestion. It may be necessary to increase the enzyme concentration in order to compensate for low DNA concentrations especially when working with bacterial samples. For partial digestion, individual restriction enzymes are titrated by adding different amounts of enzymes to DNA agarose samples incubated for only 2.5 h.

4. PFG ELECTROPHORESIS

Since PFG electrophoresis was first described (1), there have been a number of derivative techniques reported. Separation of DNA by all of the techniques is based on the same principle. DNA molecules are forced to change the direction in which they are moving. The speed at which a molecule changes direction is molecular weight dependent whereas the ordinary electrophoretic mobility of large DNA molecules is independent of molecular weight. DNA molecules are forced to change direction by exposing them

Table 8. Gel preparation and loading for the Pharmacia-LKB Pulsaphor apparatus.

A. *Preparation of one 20 cm × 20 cm × 5 mm 1% LE agarose gel*

1. Add 2.5 g of LE agarose (FMC) to 225 ml of distilled water.
2. Heat for 2 min in a microwave oven at full power.
3. Add 25 ml of 10 × modified TBE buffer, swirl and heat for a further 2 min in a microwave oven.
4. Swirl gently and inspect to make certain all agarose is in solution.
5. Cool to 50°C (i.e. until the flask is comfortable to hold in the bare hand).
6. Make sure the gel plate is on an absolutely level surface.
7. Pour 210 ml of agarose into the Pharmacia-LKB gel frame avoiding air bubbles by tilting the agarose container while pouring.
8. Remove air from the agarose feet used to hold the gel using a glass Pasteur pipette.
9. Place the comb 3 cm from the edge of the Pharmacia-LKB frame, making certain it is perpendicular to the gel plate. (Black electrical tape may be put on the back of the gel platform to mark the precise comb location and provide contrast to ease loading.)
10. Let the gel cool for at least 20 min before removing the frame. (It may be necessary to loosen the frame using a spatula.)

B. *Loading*

1. Glass rods are used to remove the inserts from Eppendorf tubes and to place the inserts on the inside of a fresh piece of parafilm. These glass rods may be conveniently prepared by melting and slightly bending the tip of a Pasteur pipette. The rod is re-usable but is alcohol-flame-sterilized between each use.
2. Use two sterile coverslips to slice each insert[a] as it rests on a clean piece of parafilm.
3. An appropriate slice of the insert is placed in the front and top of a gel well using coverslips. (Change coverslips and parafilm after each sample.)
4. Gently push the insert slice to the front and bottom of the well using a glass rod. (Avoid formation of air bubbles in between the insert slice and the running gel.)
5. Seal the inserts in the well with 0.5% low gelling temperature agarose.

[a]Liquid samples may also be loaded onto PFG gels. Such samples should be run into the agarose running gel before the buffer circulation pump is turned on (10−20 min).

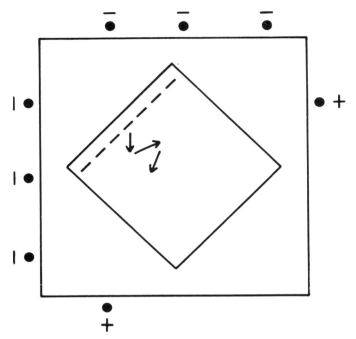

Figure 2. The electrode configuration used in a typical double inhomogeneous field PFG electrophoresis experiment.

to alternating electrical fields. The time each field is on is called the pulse time. The shift between alternate fields is essentially instantaneous. Pulse times must be tuned to the size molecules that need to be resolved (see below).

Most of our own experiments are conducted on a Pharmacia-LKB Pulsaphor apparatus. Thus, the methods described here are specifically tailored to experiments conducted on that apparatus. However, the parameters, concerns and problems are certainly encountered with other equipment.

4.1 Pouring and loading the gel

For most applications standard 1% low endoosmosis agarose in TBE is used (*Table 8*). The gel is poured on a levelling table, and care is taken to ensure that the comb is placed perpendicular to the surface of the gel. For reproducibility, the comb is always place at the same location. In the Pharmacia-LKB Pulsaphor apparatus this is 3 cm from the top of the 20 × 20 cm gel.

The best PFG resolution is obtained with low amounts of DNA. For instance, for mammalian samples we routinely load 1−3 μg per lane. One can load smaller amounts of DNA in PFG experiments because of the band-sharpening effects that result from the double inhomogeneous field configuration shown in *Figure 2*.

DNA insert samples are usually made up 4- to 10-fold more concentrated than that required per gel lane. The amount needed depends on the pulse time, the electrode configuration and the number of bands into which the DNA will be resolved. DNA amounts may be adjusted by making different size insert slices with a coverslip. This is done most conveniently by putting the DNA sample on a fresh sterile piece of parafilm. Thus, one DNA insert sample may be used in many gel lanes or may be returned to ESP for further storage at 4°C.

The gels are loaded by taking a slice of an insert sample with a coverslip and pushing it into the well. The slice should be loaded against the bottom and front of the gel. It is useful to mark above the well after the sample is loaded by cutting with a coverslip to keep track of which wells have been loaded. After all the samples are loaded they are sealed in place with liquid 0.5% low gelling temperature agarose.

4.2 Choosing effective running conditions

The usual PFG electrophoresis conditions are 3−10 V/cm, 15°C in modified TBE

Table 9. Standard PFG electrophoresis running conditions. Using 1% agarose, 12°C, modified TBE and the electrode configuration shown in *Figure 2* in a Pharmacia-LKB Pulsaphor apparatus.

Pulse time	Run time (h)	Field strength (V/cm)	Molecular size range resolved (Mb)
1 sec	18	10	0.002−0.05
5 sec	18	10	0.002−0.08
15 sec	40	10	0.05 −0.25
25 sec	40	10	0.05 −0.45
100 sec	40	10	0.05 −1.2
30 min	168	3	1.0 −2.0
60 min	168	3	1.0 −3.0
75 min	168	3	3.0 −7.0

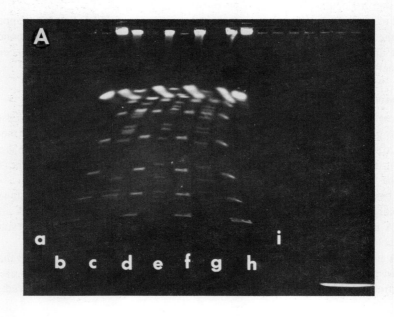

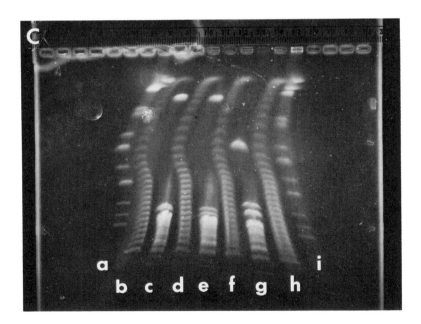

Figure 3. The effect of pulse time on the ability of PFG to resolve different size DNAs. Standard PFG running conditions were 1% agarose, 10 V/cm, 12°C, 40 h run times in modified TBE at pulse times of (**A**) 15 sec, (**B**) 45 sec and (**C**) 100 sec. The samples shown are *S.cerevisiae* chromosomal DNAs (**lanes a,i**), λ DNA concatemers (**lanes b,d,f,g**) and *Not*I-digested *E.coli* DNA (**lanes c,e,g**).

buffer. The TBE buffer is modified to reduce the conductivity by lowering the EDTA concentration (*Table 1*). Heat generated during the run would exceed the capacity of the cooling baths when normal EDTA concentrations are used. When using the electrode configuration shown in *Figure 2*, the field strength is given as the total voltage applied, divided by the shortest measurable distance between the anode and cathode electrodes. The experiments are run at 12−15°C, which means that the cooling bath must be set at 6−8°C for the Pharmacia-LKB Pulsaphor apparatus. The buffer (3 litres) is pre-cooled before starting the experiment.

For most applications the only experimental variable that must be adjusted is the pulse time. Optimum separation is obtained for those DNA molecules which spend most of a pulse time re-orienting and very little time moving, because movement is not dependent on molecular size. Resolution is based on retardation differences rather than movement of DNA. This means that long gel run times are required for the best resolution.

Typical pulse times, for resolving different sized DNA molecules on a Pharmacia-LKB Pulsaphor apparatus, are shown in *Table 9*. An extensive set of studies describe the effect of variations in field strength and shape, temperature and gel concentration (11,17−18). If these are varied it will affect the choice of pulse time in complex ways. For example the size range resolved in PFG scales roughly as the product of the square of the field strength and the pulse time (11). An example of the effect of different pulse times on the same samples is shown in *Figure 3*. Pulse times of 100 sec at 10 V/cm are extremely convenient because molecules that range in size from 0.05 Mb to 1 Mb are resolved. In order to examine molecules above 1 Mb it is necessary to lower the voltage to avoid damage to these molecules during electrophoresis.

The conditions shown in *Table 9* have been worked out for the Pharmacia-LKB Pulsaphor apparatus. They may have to be changed for other apparatus. Our standard running conditions have been chosen to maximize resolution. The lanes do not run straight. Any method that is chosen to straighten the gel lanes seems inevitably to result in a decrease in resolution. Simple ways to straighten lanes are to use short run times, use only the centre gel lanes or move the anode more towards the centre of the side of the box (see below).

The double inhomogeneous field configuration shown in *Figure 2* was found, empirically, to give the best resolution. The anode is placed at 22% of the length of the side of the box, while the cathodes are placed at 11%, 44% and 77% of the length of the side. Moving the anode towards the centre will result in less curvature and less resolution, while moving the electrode out further towards the corner will increase the curvature and band sharpening effects.

The fact that the gel lanes do not run straight does not interfere with estimating molecular sizes by comparing migration distances of samples and standards in different gel lanes (17,18). This means that mobility (and size) comparisons between gel lanes may be made by simply dropping a diagonal from the well to the band of interest. One may simply ignore the sinuosities at the top of the gel. However, the curvature at the bottom cannot be ignored. To minimize this curvature, the DNA molecules are normally run to about 5 cm above the bottom of the gels. The run times are usually 40−48 h to ensure the best resolution. Shorter runs times may also be used. They result in less resolution, straighter bands and are extremely useful for quickly evaluating an

Table 10. Common problems encountered in PFG experiments.

1. Pharmacia-LKB power supply running at constant current instead of constant voltage.

    Too much buffer used in chamber.
    EDTA concentration is too high in TBE buffer.
    10 × TBE buffer diluted incorrectly.

2. Difference in current between N/S and E/W fields greater than 30 mA.

    Box is not level. Check using a spirit level. The box may be levelled using the levelling feet on the bottom of the electrophoresis box.
    Electrodes are not firing. If there are no air bubbles around an electrode, check the platinum wire on the electrode for breaks. Under continuous use, the lifespan of the Pharmacia-LKB positive electrode is about 1 month, while that of the negative electrode is about 10 months.

3. No current indicated on Pharmacia-LKB power supply.

    Circuit is broken. Check electrode contacts and make sure all electrodes are pushed all the way down. Check copper contacts on the Pharmacia-LKB box lid.

4. Overheating of gel.

    Buffer composition is incorrect
    Circulation pump is not working. Check that levelling feet are on the electrophoresis box. If the pump is jammed, it may be easily fixed by turning the box upside down, twisting the pump off and removing any items jamming the passage.

5. No shutdown on Pharmacia-LKB power supply.

    Timer is not on ON.
    Timer switch on Pharmacia-LKB power supply should be on RUN, and ALARM switch should be on AUDIBLE WITH SHUTDOWN.

6. Samples run to one side.
    Electrical fields are not alternating.

    Check for broken or thinned platinum wire on electrodes and for good contact between electrode and guide wire.
    Check that Pharmacia-LKB CONTROL UNIT is ON.

experimental result. The minimum time we find useful is 20 h. Common problems and their causes are discussed in *Table 10* and the Appendix.

5. PHOTOGRAPHING, BLOTTING AND HYBRIDIZATION OF PFG EXPERIMENTS

Conventional photographing, blotting and hybridization protocols may be used. However, some modifications are necessary to maximize the signal and to ensure successful hybridization results with single copy mammalian probes (*Table 11*). It appears that the blotting protocol is more critical than the hybridization protocol. Large DNA must be nicked and denatured to ensure effective blotting and retention on a filter. As an alternative to blotting, it may be possible to use in-gel hybridization procedures for large DNA samples (19,20).

Pulsed-field gel electrophoresis

5.1 **Photographing and blotting**

The gels are protected from light once they are stained with ethidium bromide. Tupperware storage containers are particularly useful for staining, de-staining and hybridization experiments. We routinely transfer DNA to nitrocellulose rather than nylon filters. The signal-to-noise ratios are higher. Using nitrocellulose it is possible to detect routinely 4−6 bands of a partial digest with a single copy mammalian probe in a 2-day exposure. The most fragments we have been able to detect with such a probe is 13. With careful handling the nitrocellulose filters are re-usable at least 3−4 times. Nylon membranes have also been used for many applications. They have the potential to be re-used more times and are not as fragile as nitrocellulose. They appear to have lower

Table 11. UV blotting of PFG gels.

1.   Stain the gel with 1 μg/ml ethidium bromide on a shaker for 10 min.
2.   De-stain the gel in used TBE For 30 min.
3.   UV irradiate the gel. The time will depend on the light intensity which changes with age.
4.   Titrate your own UV box (see text).
5.   Include the time the gel was irradiated while photographing.
6.   Keep gel in the dark except when UV irradiating.
7.   Denature for 1 h with gentle shaking, using one litre of the following solutions per 400 ml gel

Final concentration	Per litre	
0.5 M NaOH	20 g NaOH	_Make fresh_
0.5 M NaCl	29.2 g	

8.   Neutralize for 1 h with gentle shaking, using one litre of the following solutions per 400 ml gel.

Final concentration	Per litre
1.5 M NaCl	87.7 g
0.5 M Tris−HCl	
(pH 7.5)	500 ml 1 M

9   Blot immediately. Blot for approximately 24 h or until 2 litres of 15 × SSC transfers (ascending transfer set up, bottom to top, is listed below). Ensure there are no air bubbles in the wells or between the gel and the nitrocellulose before proceeding.

   (i)   10 cm package of blotting paper (as a weight);
   (ii)   15 cm white paper towels;
   (iii)   4 cm stack of blotting paper (S+S, GB004);
   (iv)   two sheets of blotting paper soaked in 2 × SSC (S+S, GB002);
   (v)   nitrocellulose filter (pore size 0.45 μm)[a];
   (vi)   a puddle of 15 × SSC;
   (vii)   gel (FACE DOWN);
   (viii) two sheets (fresh) (S+S, GB002);
   (ix)   sponge[b] sitting in 15 × SSC;
   (x)   22 × 34 cm plastic Tupperware container.

10. Remove the towels carefully, turn the gel plus nitrocellulose upside down on sterile 3 mm paper and mark the wells through the gel with a marker pen. Number well #1.
11. Place the filter in 2 × SSC and gently shake the filter with 2 × SSC for 5 min to remove agarose. Inspect visually to make certain no agarose remains.
12. Air dry for 30 min. Bake in a vacuum oven at 80°C for 1 h.
13. Re-stain the gel with ethidium bromide. Take a photograph of the stained gel to determine how much DNA is left in the gel. About 20% of the DNA should _remain_ in the gel.

[a]Prepare the nitrocellulose by soaking in double-distilled water until saturated; then soaking in 2 × SSC.
[b]5 cm thick sponges should be rinsed extensively in distilled water before use.

58

signal-to-noise ratios. The same blotting protocol should be used for both nitrocellulose and nylon. Movement of the DNA to the membrane is mediated by the buffer that travels through the gel. Thus, routinely 2 litres of buffer is allowed to transfer. NaOH transfer procedures onto nylon membranes have been used successfully in some experiments (21). Recently, preliminary experiments with the Pharmacia-LKB Vacublot showed efficient (2 h) transfer of megabase DNA. That is, single copy mammalian sequences were detected in hybridization experiments in 1-day exposure. This is the same exposure time that is required when the capillary transfer procedure is used.

DNA can be nicked prior to transfer either with acid (HCl) or by UV irradiation in the presence of ethidium bromide. We find that large DNA molecules are much more sensitive to nicking by HCl than small molecules. Thus, while it is possible to nick megabase DNA with acid, careful titration and monitoring are required to ensure reproducibility. Instead, UV nicking appears to be much more forgiving. However, each UV source must be titrated for the exposure which gives the best hybridization signal. Also one should bear in mind that the light intensity will decrease with use. Both 245 nm and 300 nm light sources may be used. A 30-sec exposure is required with a 300 nm light source with an intensity of approximately 5.3 μW/cm as measured with a UVX31 Radiometer. However, light intensity meters are usually not accurate and can vary by 90% even with the same meter, whether or not it has been recently standardized. Thus, we routinely evaluate the transfer by re-staining the gel after transfer.

The following rules of thumb may be used to estimate trial UV doses using fast film (Polaroid 667):

(i)   20 sec of exposure to a light source in which a photograph can be taken in less than 1 sec;

(ii)   10 min of exposure for an old or weak light source with which 1.5 min of exposure are required for photographing.

All exposure to UV light is counted. Thus any desired gel photographs are taken during the nicking procedure.

5.2 Hybridization

There have been many different protocols used, and it is not clear at this time which is best. Hybridization has been done successfully with and without formamide using a variety of reagents and formulations that are much too numerous to elaborate. One hybridization mix used containing dextran sulphate is shown in *Table 1*. This solution is used both for a 1-h pre-hybridization treatment and the hybridization itself.

Nick-translated or oligonucleotide-labelled (22,23) probes may be used. Routinely we make up our own reagents and obtain specific activities of $1-3 \times 10^9$ c.p.m./μg using the latter labelling procedure. A maximum of $2-4$ ng of probe per ml is added when using a hybridization mix containing dextran sulphate. One may include a total of 5×10^4 c.p.m. of [^{32}P]λ DNA to label λ size standard during hybridization. The λ DNA does not appear to cross-hybridize with human genomic DNA.

Normal high stringency washes may be used. However, when getting started it is sometimes useful to wash the blot with much lower stringencies. For instance, in some mammalian samples, single fragment signals were detected with single copy probes using an outrageously low stringency of only 2 × standard saline citrate (SSC) at 50°C.

It may also be useful to include in the PFG experiments a restriction enzyme digestion with a known result. This can either be run in the PFG electrophoretic experiment or added later by including a strip from an ordinary electrophoresis in the hybridization bag. Examples of bad gel runs and a discussion of their uses are shown in the Appendix.

6. ACKNOWLEDGEMENTS

The authors would like to thank the many researchers for their comments and suggestions, including Simon Lawrance, Peter Warburton, Eric Schon, Marthe-Elisabeth Eladare, Osami Niwa, Mitsuhiro Yanagida, Andras Gaal and Fred Alt. This work was supported by grants from the NIH, GM14825, the NCI, CA39782, the DOE DE-FG02-87ER-GD852; the Hereditary Disease Foundation, the MacArthur Foundation and Pharmacia-LKB.

7. REFERENCES

1. Schwartz,D.C., Saffran,W., Welsh,J., Haas,R., Goldenberg,M. and Cantor,C.R. (1983) *Cold Spring Harbor Symp. Quant. Biol.*, **47**, 18.
2. Schwartz,D.C. and Cantor,C.R. (1984) *Cell*, **37**, 76.
3. Van der Ploeg,L.H.T., Schwartz,D.C., Cantor,C.R. and Borst,P. (1984) *Cell*, **37**, 77.
4. Smith,C.L., Warburton,P.E., Gaal,A. and Cantor,C.R. (1986) In *Genetic Engineering*. Setlow,J. and Hollaender,A. (eds), Plenum Press, New York, Vol. 8, p. 45.
5. Smith,C.L. and Cantor,C.R. (1986) *Cold Spring Harbor Symp. Quant. Biol.*, **51**, 115.
6. Smith,C.L. and Cantor,C.R. (1987) In *Methods in Enzymology*. Wu,R. (ed.), Academic Press, San Diego, Vol. 155, p. 449.
7. Smith,C.L., Lawrance,S.K., Gillispie,G.A., Cantor,C.R., Weissman,S.M. and Collins,F.S. (1987) In *Methods in Enzymology*. Gottesman,M. (ed.), Academic Press, San Diego, Vol. 151, p. 461.
8. Smith,C.L., Econome,J.G., Schutt,A., Klco,S. and Cantor,C.R. (1987) *Science*, **236**, 1448.
9. Lawrance,S.K., Smith,C.L., Weissman,S.M. and Cantor,C.R. (1986) *Science*, **235**, 1387.
10. Cantor,C.R., Mathew,M.K. and Smith,C.L. (1988) *Annu. Rev. Biophys. Chem.*, **17**, 287.
11. Mathew,M.K., Smith,C.L. and Cantor,C.R. (1988) *Biochemistry*, in press.
12. Smith,C.L. and Cantor,C.R. (1987) *Trends Biochem. Sci.*, **12**, 284.
13. Cook,P.R. (1984) *EMBO J.*, **3**, 1837.
14. Fairlamb,A.H., Weisloge,P.O., Hoeijmakers,J.H.J. and Borst,P. (1978) *J. Cell Biol.*, **76**, 293.
15. Smith,C.L. and Kolodner,R. (1988) *Genetics*, **119**, 227
16. Larsen,A. and Weintraub,H. (1982) *Cell*, **29**, 609.
17. Mathew,M.K., Smith,C.L. and Cantor,C.R. (1988) *Biochemistry*, in press.
18. Cantor,C.R, Gaal,A. and Smith,C.L. (1988) *Biochemistry*, in press.
19. Rao,R.V., Labie,D. and Krishnamoorthy,R. (1987) *Nucleic Acids Res.*, **15**, 4355.
20. Purrello,M. and Palazs,I. (1983) *Anal. Biochem.*, **128**, 393.
21. Drum,M.L., Smith,C.L., Dean,M., Cole,J.L., Iannuzzi,M.C. and Collins,F.S. (1988) *Genomics*, in press.
22. Feinberg,A.P. and Vogelstein,B. (1983) *Anal. Biochem.*, **132**, 6.
23. Feinberg,A.P. and Vogelstein,B. (1983) *Anal. Biochem.*, **137**, 266.

8. APPENDIX

This section contains the types of PFG gel photographs that are not usually published. They are shown here as examples of particular problems to aid experimenters in diagnosing their own problems. In most of the gels, the outside lanes contain *S.cerevisiae* chromosomal DNAs and the next inner lanes contain concatemerized λ bacteriophage DNA size standards. The inner lanes contain restriction enzyme-digested bacterial or mammalian DNA. The gel runs were intended to be carried out under standard conditions at 25 or 100 sec pulse times.

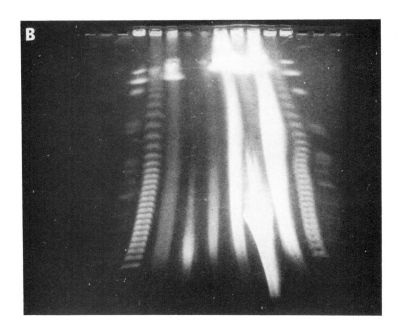

Figure 1A and **B.**

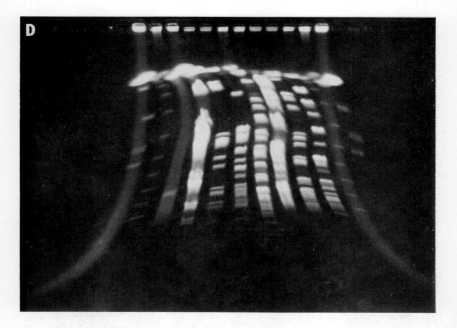

Figure 1. An example of overloading of DNA samples. Most of the size standards loaded are fine. However, most of the mammalian (**A**−**C**) and bacterial samples (**D**) are overloaded.

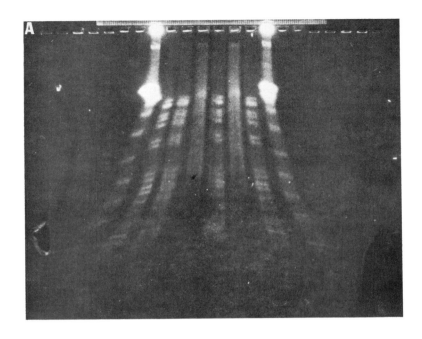

Figure 2A and **B.**

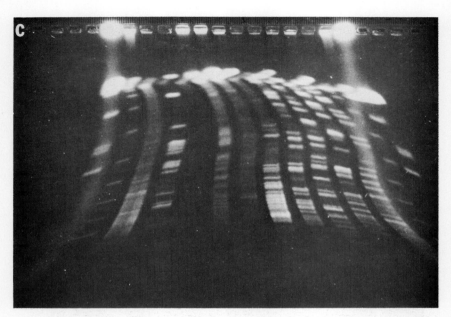

Figure 2. An example of improper preparation of DNA samples. (**A**) Protein is still bound to some of the bacterial DNA samples. (**B**) Diffusible material, probably protein, has not been dialysed away from the *S.cerevisiae* DNA samples. (**C**) Restriction enzyme digestion of bacterial DNA samples is incomplete probably because proteinase K was not completely inactivated or samples were not incubated for a sufficient time in ESP prior to restriction enzyme digestion.

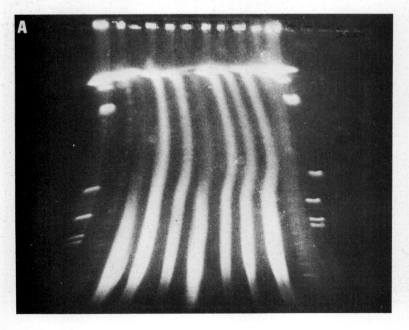

Figure 3A−C.

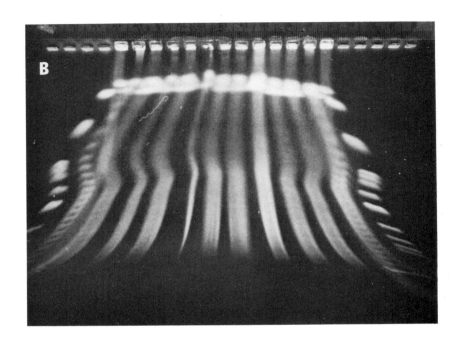

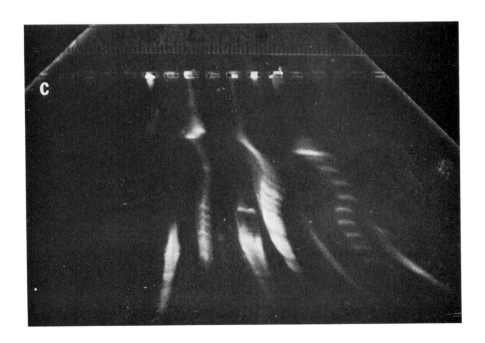

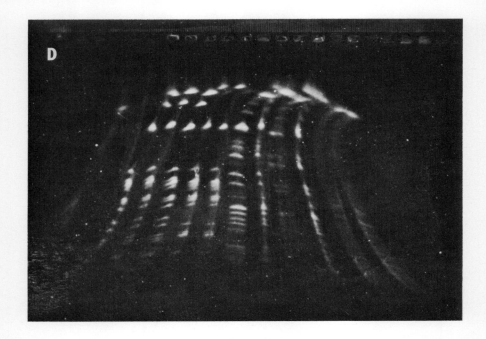

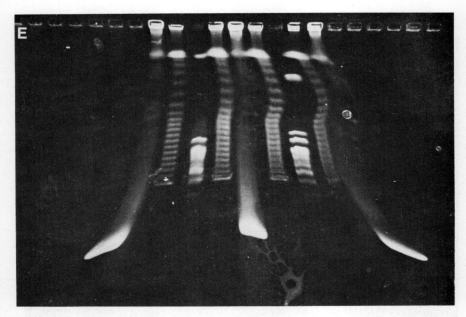

Figure 3. An example of improper loading of DNA samples. (**A** and **B**) Some, but not all, of the mammalian DNA sample inserts were not loaded flat against the front of the wall. (**C** and **D**) All of the bacterial DNA samples were loaded unevenly. (**E**) Bacterial and λ phage DNA samples were loaded well, but *S. cerevisiae* samples were not.

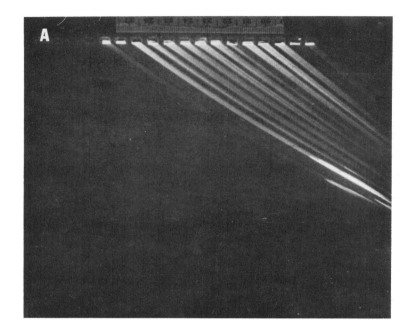

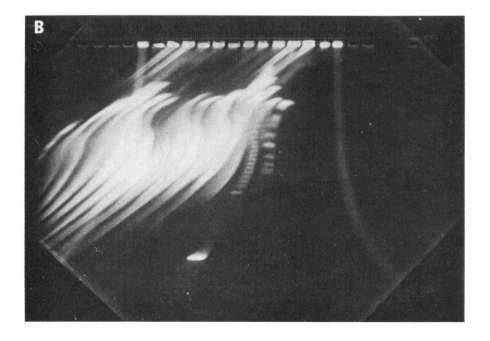

Figure 4A and **B.**

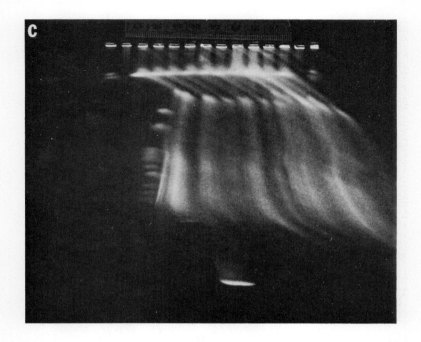

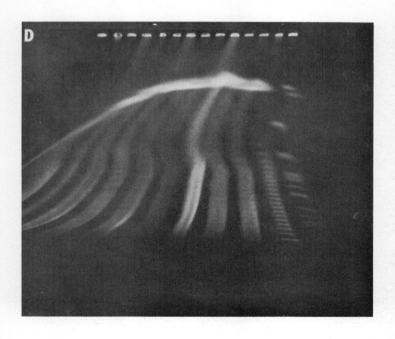

Figure 4. An example of uneven pulsing. (**A**) No alternate pulsing. (**B** and **C**) Anode burned out during run (samples are also overloaded in **B**). (**D** and **E**) One of the cathodes was not connected.

Figure 5A.

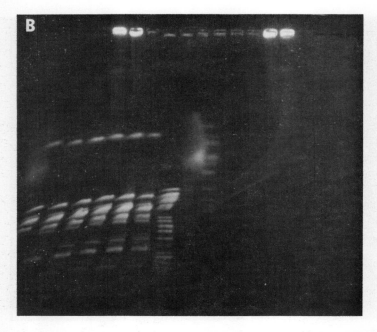

Figure 5. An example of uneven electrical fields because of improper levelling of the PFG electrophoresis box. (**A**) Mammalian (a few samples are slightly overloaded) and (**B**) bacterial DNA samples.

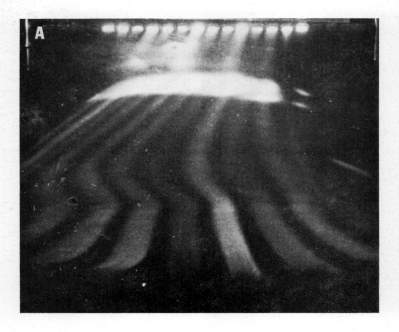

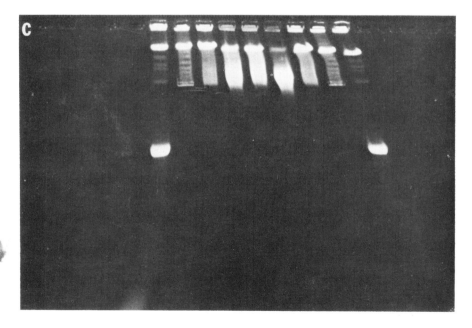

Figure 6. An example of improper run times. (**A** and **B**) Although the run times were correct, the samples have run too far because the temperature was higher than usual. (**C**) Run time was too short.

Figure 7. This is a test for the reader.

CHAPTER 4

Chromosome jumping

FRANCIS S.COLLINS

1. INTRODUCTION

The last decade has witnessed a remarkable series of stunning successes in the area of molecular biology. The development of standard tools for the reliable cloning, sequencing and expression analysis of eukaryotic genes has rapidly advanced our understanding of gene structure and regulation, and has permitted the molecular definition of the specific genetic abnormality causing a variety of human diseases. At the same time, progress in mapping genes to specific parts of human chromosomes has also proceeded rapidly. At the recent Human Gene Mapping workshop in Paris (HGM9, September 1987) a total of 1360 genes and DNA segments were reported to be mapped to specific chromosomes. The accumulation of mapping information continues to proceed at an exponential rate. However, as shown in *Figure 1*, the molecular cloning and gene mapping areas of endeavour operate on quite different scales.

Until recently, therefore, it has not been clear how information in the mapping area might be able to guide efforts in molecular cloning. The 'resolution gap' in the 100 – 5000 kb range, within which, until recently, no convenient molecular techniques existed, contributed greatly to the difficulty in proceeding from mapping information, derived by somatic cell genetics or linkage analysis, to the precise molecular definition of a genetic disorder. This process is now commonly but perhaps somewhat inappropriately referred to as 'reverse genetics' (1,2).

Fortunately, in the last few years several approaches have become available which operate in the size range necessary to deal with this 'resolution gap'. In this chapter, the technique of chromosome jumping (3 – 5), which is able to provide DNA clones at a substantial distance from an original probe, will be described. Protocols for the construction of chromosome jumping libraries and linking clone libraries (4,5) will be outlined, and the advantages and disadvantages of this approach will be discussed. Section 6 deals with potential refinements of the method which promise to increase the distances approachable by this technique, but which are still in the development phase.

2. SCALE CONSIDERATIONS

In *Figure 2*, the size ranges over which a variety of current genetic techniques operate is schematically depicted. The y axis is a logarithmic scale of physical distance, marked both in base pairs and in kilobases (1 kb = 1000 bp). Also shown on this scale is a genetic distance indicator, using the relationship that one centiMorgan (cM) is approximately 10^6 bp. This relationship must be considered imperfect, since the relationship between genetic and physical distance is non-linear along a chromosome

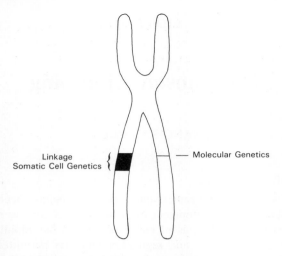

Figure 1. Schematic depiction of the differing scales approachable by molecular cloning on the one hand, and gene mapping on the other.

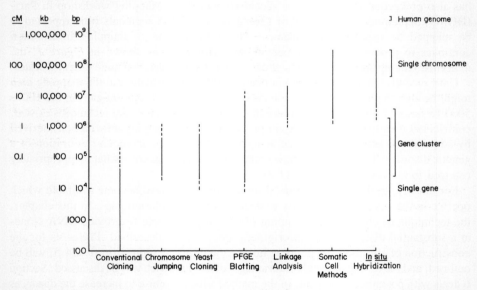

Figure 2. Quantitation of the size range over which a variety of techniques are currently operational. The scale on the vertical axis is marked in physical distances (bp and kb), and also in genetic distances (cM), using the approximate relationship that 1 cM = 1000 kb.

and can be affected by the presence of recombination 'hot spots'. Such hot spots may lead to the situation where a relatively large genetic distance is represented by a small physical distance. On the other hand, there may well be regions of the genome where recombination is unlikely, which will lead to the reverse situation. As shown in *Figure 2*, classical molecular genetic techniques operate well up to the 50 kb region, which is the maximum insert size clonable in a cosmid vector. Distances larger than this can be cloned by chromosome walking, in which one uses the end of one clone to re-screen a library and obtain an overlapping clone; regions as large as several hundred kilobases have been obtained by this approach. However, this can be quite time consuming and one can encounter regions which are impossible to cross, either because of a long stretch of repetitive sequences or because of the presence of a sequence of DNA which is unclonable in standard vectors. If the gene of interest is known to be several hundred kilobases away, chromosome walking can be problematic.

At the other end of the spectrum, somatic cell genetics, *in situ* hybridization and genetic linkage analysis rarely give a resolution better than 1000−5000 kb. The three approaches noted in the middle of the figure, namely pulsed-field gel electrophoresis (6−11), chromosome jumping (3−5) and yeast cloning (12), have provided a means of working with large molecular distances and allowing the use of mapping information to direct a search toward specific molecular lesions.

The importance of these approaches is in allowing access to molecular understanding of a large number of disorders for which the function of the responsible gene is completely elusive. A great number of human single gene disorders, such as cystic fibrosis, Huntington disease and neurofibromatosis, are known to be inherited in a strict Mendelian fashion and have a phenotype which is well characterized on a descriptive level, but the normal function of the responsible gene has not been defined. The advent of linkage analysis using polymorphic DNA markers (restriction fragment length polymorphisms, RFLPs), has allowed the mapping of genes for such disorders to specific human chromosomes (13,14), which has in turn directly led to an increased ability to perform pre-natal and pre-symptomatic diagnosis. However, in order to use such mapping information to actually clone the responsible genes and obtain an understanding of the biology which might lead to therapeutic interventions, techniques operational in this region of large molecular distances are required. Chromosome jumping is such a technique; its principle and a few applications will be described in the subsequent sections.

3. GENERAL JUMPING LIBRARIES

3.1 Types of jumping

In this chapter we will draw a distinction between general jumping libraries (3−5,15), which are constructed in such a way that they allow one to start at essentially any location in the genome and travel a specified distance along the chromosome, and specific jumping libraries (4,5,16), which consist of clones which jump from a relatively rare restriction site, such as *Not*I, to the next adjacent such restriction site. These types of jumping are schematically represented in *Figure 3*. It should be apparent that the construction and use of these libraries is somewhat different. General jumping libraries, because they must represent all of the sequences in a genome to be useful, are technically more difficult to construct. In general, $1-3 \times 10^6$ clones are needed in such a library in order to have a high likelihood of being able to use such a library from any given start point. Specific jumping libraries, on the other hand, may only need 10 000−20 000 clones to be complete, since the number of independent clones in such a library is equivalent to the number of restriction fragments generated by the enzyme being used. For an enzyme such as *Not*I which cuts approximately every 1000 kb in the human

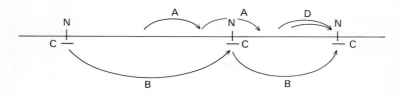

Figure 3. Scheme of the various types of jumping libraries. **A** depicts general jumping library clones, which allow one to jump from any starting point a specified distance along a chromosome. In **B** are two specific jumping library clones, in this instance for the enzyme *Not*I (N), which jump from one such site to the nearest adjacent one. **C** denotes linking library clones, which are contiguous DNA fragments including a specific rare restriction site, in this case *Not*I. **D** denotes jumping library clones which begin in a random location and jump to the nearest adjacent rare restriction site, again in this example *Not*I.

genome, there should only be approximately 3000 such fragments; therefore a library of 10 000 *Not*I jumping clones is likely to be essentially complete. The obvious disadvantage of specific jumping libraries is that they cannot be used when the start point of a jump is not located adjacent to a rare restriction site. Unfortunately, this is often the case, which has limited the rapid application of this approach. However, specific jumping libraries can be of great utility, once a clone adjacent to a rare restriction site is identified, to allow one to start using such a library.

A third type of library, which connects rare restriction sites to adjacent random genomic sequences, would in theory solve this problem by allowing one to travel from any initial start point to the nearest rare restriction site, which could then be used as a means of entry into a specific jumping library. Unfortunately, however, this third type of library has been thus far very difficult to construct. Such a library, like the general jumping library, would have to contain several million clones, including jumps of widely varying sizes. This approach will not be further discussed here.

3.2 Principle of the method—general jumping libraries

The basic strategy of chromosome jumping is to circularize very large DNA fragments by ligating them in dilute solution. The junctions of such circles bring together fragments of DNA which were located at large distances apart in the genome. Selective cloning of these junction fragments in standard vectors then allows the generation of a library of jumping clones. This strategy, as applied to the construction of a general jumping library, is diagrammed in *Figure 4*.

There are several important considerations before beginning the construction of the library

(i)   What size of jump is desired? This will be determined by the size selection of the partially digested DNA. As further described below, the difficulty in constructing the jumping library increases as the 3/2 power of the jump size.

(ii) What enzyme to use? Ideally, one would want a completely random collection of DNA fragments, in order to have a high likelihood that all genomic sequences will be represented at the ends. In practice (15), we have used *Mbo*I (or its isoschizomer *Sau*3A) since these enzymes cut approximately every 250 bp. Stretches of DNA which are devoid of such sites, however, will not be represented in the library.

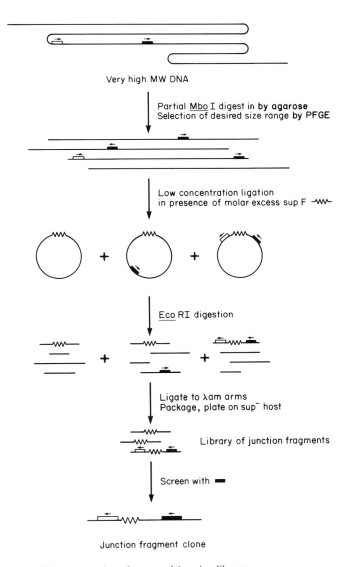

Very high MW DNA

Partial <u>Mbo</u> I digest in **by agarose**
Selection of desired size range **by PFGE**

Low concentration ligation
in presence of molar excess sup F

<u>Eco</u> RI digestion

Ligate to λam arms
Package, plate on sup⁻ host

Library of junction fragments

Screen with ▬

Junction fragment clone

Figure 4. Principle of the construction of a general jumping library.

(iii)   What source of DNA to use? For optimum utilization of the library for a variety
of starting probes, it may be advantageous to use a source of DNA which
represents the whole genome of the organism, for example peripheral lymphocytes
from a blood donor for a human library. However, if one has a specific application
in mind which involves a specific chromosome, there may be some advantage
in using a somatic cell hybrid which contains only that chromosome on a
background of chromosomes of another species. The advantage of such a strategy
is that it allows the immediate testing of any jumps obtained from the library

for validity, by simply checking the jumping fragment obtained for its correct species mapping (17).

3.3 Preparation and size selection of DNA

The DNA to be partially digested must be of very high molecular weight (> 1000 kb). The protocol which we follow is basically the same as that used to prepare DNA for pulsed-field gel electrophoresis (10,11), except that the amount of DNA is scaled up for preparative purposes.

(i) Using the fact that one mammalian cell contains 6.7 pg of DNA, grow a sufficient number of tissue culture cells to represent the desired amount of DNA. A typical quantity of DNA needed for construction of a jumping library is 200 μg, which represents approximately 3×10^7 cells. Because the quality of the DNA must be high, it is very important that the cells be regularly fed and carefully examined prior to harvesting. Alternately, one can prepare cells from peripheral blood by Ficoll−hypaque centrifugation (18).

(ii) After harvesting, count the cells accurately in a haemocytometer and suspend in a sufficient volume of phosphate-buffered saline (PBS) so that the concentration of cells is approximately 2×10^7/ml. Then mix quickly with an equal volume of 2% low melting temperature agarose (Seaplaque, FMC) which is 125 mM in EDTA and has been melted and kept at a temperature of 40°C. Pipette into a lucite mould; in addition to the usual size blocks, the mould should include wells which are $2 \times 8 \times 135$ mm and will serve as a source of preparative-scale DNA. In general, it is helpful to make 10−20 standard DNA blocks which can be used for test digestions, and two large blocks which will be used for the actual scale up.

(iii) Prepare the DNA by proteinase K digestion in the presence of large quantities of EDTA and *N*-laurylsarcosine as previously described (10,11,15).

(iv) After preparation, test the DNA for its integrity and the absence of nucleases. Incubate half a block at 37°C for 3 h in the presence of 10 mM $MgCl_2$. Load this half block and another untreated half onto a field inversion (8) or OFAGE (7) gel and analyse the DNA size. To be useful for chromosome jumping, the DNA in the undigested block should be nearly all retained in the well, and the block which has been incubated in the presence of $MgCl_2$ should not show any degradation. If degradation is apparent with magnesium, this indicates that there is still nuclease contamination in the blocks, and they must then be re-cycled through proteinase K. This is usually successful in eliminating this problem.

(v) Some sources of DNA routinely give a detectable amount of degraded DNA in the 50−100 kb size range, which probably arises from dead cells. This is particularly true for lymphoblasts. Because this degraded DNA can be quite damaging to the jumping protocol, it is useful to eliminate this material from the agarose blocks prior to restriction enzyme treatment. This is done by placing either small blocks or preparative scale blocks in the wells of an OFAGE gel and running the gel with a 20 sec pulse interval for approximately 2 or 3 h. This will electrophorese DNA molecules smaller than about 100 kb out of the blocks but leave the high molecular weight DNA essentially unaltered. The blocks can

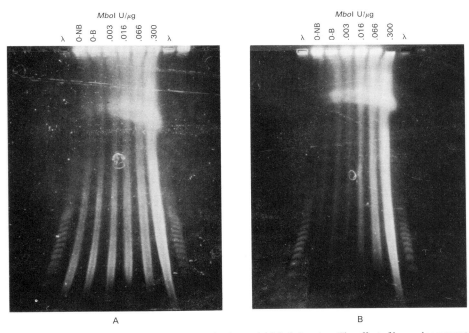

Figure 5. Preparing DNA for chromosome jumping by partial *Mbo*I digestion. The effect of increasing amounts of enzyme on the range of DNA produced is shown. In addition, in (**A**) the agarose blocks containing DNA have been used directly, whereas in (**B**) a pre-electrophoresis step has been carried out on these blocks to eliminate smaller DNA molecules prior to the *Mbo*I digestion. The smaller molecules can interfere with jumping libraries, and are effectively eliminated by this step. In each instance, size markers, consisting of multimers of lambda (genome size 48.5 kb) are included.

> then be removed from the gel, and represent a more reliable source material.

(vi)   Carry out test digestions on half blocks using a range of concentrations of *Mbo*I. Digest the blocks in Eppendorf tubes in a volume of 300 μl and the standard restriction enzyme buffer, using amounts of enzyme from 0.03 units to 0.15 units. Since each half block contains approximately 3.3 μg of DNA when made by the above protocol, this represents a range of enzyme concentrations from 0.01 U/μg to 0.045 U/μg. Carry out the digestions at 37°C for 1 h, and stop the reaction by the addition of 10 μl of 0.5 M EDTA. Electrophorese the test digestions on a pulsed-field gel to determine the optimum concentration for generating the size of molecules desired. An example is shown in *Figure 5*, which also demonstrates the utility of the pre-electrophoresis step to eliminate small amounts of low molecular weight degraded DNA prior to restriction digestion.

(vii) Scale up the concentration of enzyme giving optimum amounts of DNA in the desired size range, using one of the preparative-scale agarose blocks and keeping the concentration of enzyme in U/μg and the DNA concentration in μg/ml the same.

(viii) Place the digested preparative scale block at the origin of a pulsed-field gel and electrophorese it using a pulse interval which gives good resolution of the size range desired, and including appropriate size markers. If an OFAGE gel (7) is used, it is important to include size markers at both edges and in the middle of the gel so that a good interpolation of the desired size range can be carried out. We have found it preferable to use a field inversion (8) or CHEF (9) gel for the scaled up electrophoresis, so that the desired range of DNA molecules is linear rather than parabolic.

(ix)   It may be important not to expose the DNA to be used for cloning to UV radiation. Therefore, cut off the ends of the gel containing the size markers and a small amount of the digested genomic DNA, stain these with ethidium bromide for UV visualization, and then use this information to identify the desired size region of the gel.

(x)   Cut the desired size range of DNA from the gel. In this and succeeding steps, it is important to use materials which are nuclease free. To release the DNA from the gel for circularization, there are two possible approaches. We have usually found it convenient to do this by electroelution. To do this, place the gel block containing the DNA into a dialysis bag, surround the gel block with about four volumes of $0.5 \times$ TBE, which is used as the running buffer for the preparative gel, and then place the dialysis bag in a horizontal electrophoresis box. Electroelute the DNA out of the gel for about 2 h using a constant forward voltage of 100 V and $0.5 \times$ TBE running buffer. The dialysis bag can then be unclamped and the gel fragment removed and stained with ethidium bromide to be certain that all the DNA has been electroeluted. After removal of the gel block, re-clamp the dialysis bag and dialyse the DNA with several changes of 10 mM Tris pH 7.4, 1 mM EDTA(TE) in order to prepare it for ligation.

(xi)   Alternatively, the DNA can be prepared by using LMT agarose for the preparative size-selection gel, cutting the desired size range out of this gel, and then melting the gel at 65°C. After melting the DNA can be diluted to the desired concentration and ligation carried out in the presence of agarose without significant inhibition of the reaction. At the time of writing, there is no strong reason to prefer one method of DNA recovery over the other. It is crucial, however, to avoid all manipulations which might shear the DNA after it is removed from the gel. Specifically, centrifugation, vortexing and pipetting must not be carried out, and it is helpful to proceed with ligation as rapidly as possible.

3.4 Circularization

The success or failure of the chromosome jumping approach depends crucially on the ability to ligate large DNA molecules into circles. Consideration of the strategy diagrammed in *Figure 4* makes it clear that tandem ligation of unrelated DNA molecules will result in the generation of junction fragments which connect two unrelated DNA sequences. Such junction fragments will result in anomalous clones which cause one to jump from a given starting point to a random sequence in the genome. To minimize this risk, it is essential to carry out the ligation at sufficiently low concentration so that on averge less than $5-10\%$ of the ligations will be tandem.

The theory of ligation of DNA molecules in dilute concentration was worked out in the 1950s by Jacobson and Stockmayer (19). The predictions of this theory have been borne out experimentally for several sizes of DNA. A very useful relation is the following:

$$j = \left(\frac{3}{2\pi\, lb}\right)^{3/2} \text{ends/ml}$$

In this equation, j is the concentration at which the ligation of a DNA molecule of segment length b and total contour length l is equally likely to occur by an inter- or intramolecular event. Obviously, in order to favour circularization, one would like to work at a concentration much lower than j. By substituting the known parameters of DNA in aqueous solution, this equation can be reduced to:

$$j = \frac{63.4}{(\text{kb})^{1/2}} \ \mu\text{g/ml}$$

where kb is the length of the DNA in kilobases (3). At a given DNA concentration i, the proportion of ligations which will be circular is $j/(i+j)$, so that in order to have 90% circular ligations, one should work at a concentration given by:

$$i_{90} = \frac{7.0}{(\text{kb})^{1/2}} \ \mu\text{g/ml}$$

In general, to produce $1-3 \times 10^6$ jumping clones in the final library, it is necessary to have approximately 0.5 μg of junction fragments after the circular molecules are digested. Since the average junction fragment is about 5 kb, for 100-kb jumps this would translate into 10 μg of size-selected DNA, whereas for 200-kb jumps 20 μg would be needed. Thus, two factors go into the ligation volume: (i) the amount of DNA necessary in order to obtain a sufficient number of junction fragments to generate the library, which goes up linearly with the size of the jump; (ii) the need to favour circular ligation, which goes up as the square root of the length of the DNA. In effect, therefore, ligation volume must be increased as the 3/2 power of the length of DNA as the jump size is increased. Typical parameters for generation of a library are shown in *Table 1*, which is for a 100-kb jump size (20).

Table 1. Parameters for 100-kb hopping library.

Starting amount of high mol. wt genomic DNA	100 μg
Amount of size-selected DNA	5 μg
Range of sizes included	80−130 kb
Amount of *supF* gene (*Bam*HI ends)	2 μg
Molar excess of *supF* gene (220 bp)	200:1
Ligation volume	25 ml
Genomic DNA concentration	0.2 μg/ml
Amount of λ vector	150 μg
Total plaques on sup$^+$ host	4×10^8
Insert containing plaques on sup$^+$ host	5×10^7
Total plaques on sup$^-$ host	2×10^6

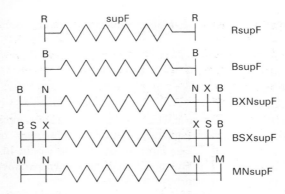

Figure 6. A variety of suppressor tRNA (*supF*) genes which have been prepared for use as selectable markers in the construction of jumping and linking libraries. To increase their utility, a variety of restriction sites have been added to the ends of this 200-bp gene, beginning with the original *Eco*RI-flanked *supF* gene described by Dunn (21). The other *supF* genes were constructed using a variety of linkers. R = *Eco*RI, B = *Bam*HI, X = *Xho*I, N = *Not*I, S = *Sfi*I, M = *Mlu*I.

It is, of course, essential to mark the junctions of the circles so that they can be selectively cloned in succeeding steps. We have used a suppressor tRNA (*supF*) gene as a biological selection (3,15), although other possible selections are possible (see below). The suppressor gene must, of course, have ends compatible with the *Mbo*I ends of the genomic DNA. Several different *Bam*HI-ended *supF* genes have been generated for this purpose and are shown in *Figure 6*. The presence of additional restriction sites internal to the *Bam*HI site is useful for subsequent analysis of the jumping clones.

The actual protocol for circularization is as follows.

(i)   Using the equations above, dilute the DNA to the desired concentration in 50 mM Tris pH 7.4, 1 mM EDTA.

(ii)   Add a 100- to 500-fold molar excess of purified *Bam*HI-ended *supF*, which has been pre-tested for ligation efficiency by self-ligation and subsequent gel analysis. We find it most convenient to prepare the *supF* by gel electrophoresis and electroelution. It is important that this fragment be as plasmid-free as possible, as small amounts of plasmid are likely to show up in the final library as clones which hybridize with plasmid-bearing probes. Allow the *supF* and genomic DNA to diffuse together for 30 min.

(iii) Bring the magnesium concentration to 10 mM and allow to equilibrate for 10 min. Then add T4 DNA ligase to a final concentration of 1−2 units/μl. Ligate for 12 h at 14°C, and then add a second aliquot of ligase and ligate for an additional 12 h.

(iv)   Recover the circularized DNA by adding 20 μg of yeast tRNA as carrier and then ethanol precipitating, spinning the pellet at 23 000 r.p.m. in a SW27 ultracentrifuge rotor. Resuspend the pellet in 100 μl of TE, and then block any unligated ends by treating with alkaline phosphatase, or filling in with Klenow. This step is important, since if ligation was incomplete in the low concentration step, unrelated fragments may generate anomalous junction fragments in the ligation to vector [Section 3.5(iii)]. Phenol extract and then precipitate the DNA and digest with *Eco*RI.

(v) A very useful way to test the success of the preceding steps is to save small aliquots (2−5%) of each step, and then run these aliquots together on a 1.4% agarose gel, followed by transfer to nitrocellulose or nylon and blotting with the *supF* gene. In a successful reaction, the *supF* gene should be seen to form a ladder during the circularization ligation, with a small amount of *supF* appearing in the very high molecular weight range of the gel indicating its ligation to genomic DNA. After *Eco*RI digestion, the ladder of *supF* should remain but the high molecular weight *supF* should be reduced to a smear of fragments in the 1−20 kb range.

(vi) Phenol extract and ethanol precipitate the *Eco*RI-digested DNA.

3.5 Cloning and screening

At this point, the *Eco*RI-digested genomic circles are ready for ligation and selection of the junction fragments. This selection is carried out by ligating the genomic fragments into a phage vector which bears amber mutations in at least two of its coat protein genes, packaging the DNA *in vitro*, and then plating on a bacterial host which lacks *supF* and will, therefore, only allow phage genomes which carry their own *supF* gene to replicate and form plaques.

In theory, any phage vector bearing amber mutations and an *Eco*RI cloning site should be suitable. One would like the cloning capacity to be as large as possible, in order not to discriminate against relatively large *Eco*RI fragments. In the best of all worlds the genomic circles would be digested only partially with a frequent cutter, and then ligated into a large capacity vector so that no such discrimination by the location of restriction sites would occur. However, the necessary losses which result from partial digestion turn out to be prohibitive if one is trying to generate a complete library. Thus far, in our hands all general jumping libraries have been complete digests of the genomic circles with *Eco*RI. Other restriction enzymes for which amber-mutated phage cloning vectors are available, such as *Hin*dIII, can also be used. In fact, such *Hin*dIII/ *Mbo*I libraries are a useful complement to the *Eco*RI/*Mbo*I libraries described here, as will be alluded to below.

In practice, we have found the use of a Charon3A vector, modified in Dr Frederick Blattner's laboratory to convert it into a zero-insert *Eco*RI vector, to be far superior to any other such vector, including Charon16A. *Figure 7* shows a map of the modified Charon3A vector, which is called λCh3AΔlac. It has a genome size of 38.5 kb, grows vigorously and gives high yields, and can accept inserts from 0 to 12 kb.

The actual protocol for generating the library is given below.

(i) Phenol extract and re-precipitate the *Eco*RI-digested genomic DNA, resuspending it in as small a volume as possible.

(ii) Digest the cloning vector with *Eco*RI, and then remove the enzyme by phenol extraction with re-precipitation. The completeness of *Eco*RI digestion, as well as the ligatability of the *Eco*RI sites, should be checked by re-ligating a small quantity of this and comparing the packaging efficiency of uncut vector, *Eco*RI-cut vector and *Eco*RI-cut and re-ligated vector.

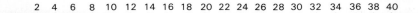

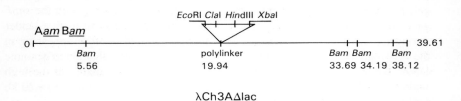

Figure 7. Restriction map of the λCh3AΔlac cloning vector. This was prepared by Dr Frederick Blattner's laboratory by removal of the 8.6-kb *Eco*RI − *Bam*HI fragment from λCh3A and replacement by the *Eco*RI − *Bgl*II polylinker fragment from the miniplasmid πVX. The resultant vector has a genome size of 39.6 kb and can be used to clone *Eco*RI fragments from 0 to 12 kb.

(iii)   Ligate the vector to the genomic insert, using a 4:1 molar excess of vector in order to cut down on the number of clones which have multiple inserts. Such clones will not, however, be misleading in the long run, since the presence of two *Eco*RI fragments in a clone is diagnostic of this event, and therefore such clones are readily recognizable.

(iv)   Test package 0.25 μg of this ligated material. Because of the large amount of vector which has to be ligated to generate a library (see for example *Table 1*), it is impractical to use commercially available packaging extracts. We use a standard two-component packaging extract as described by Hohn (22). In order for the efficiencies to be high enough to generate a complete jumping library, it is crucial that the extracts give an efficiency of $3-4 \times 10^8$ p.f.u./μg of wild-type λ DNA when tested.

(v)   If the efficiency of the jumping library is close to the desired level but not quite high enough, an additional factor of 3 can usually be achieved by adding a crude preparation of lambda terminase to the packaging reaction. Terminase is a phage protein which cleaves ligated DNA at the *cos* site after packaging into the phage head, and is usually the limiting protein for packaging efficiency in extracts made by the protocol referenced above. A crude preparation of terminase can be conveniently prepared by using the strain AZ1069, prepared by Dr H.Murialdo, which contains the terminase gene on a selectable plasmid under temperature-inducible control (23). Shifting the temperature of a culture upward at the appropriate log phase of growth results in the production of high levels of terminase which can be readily prepared simply by sonicating the culture. The activity of the terminase can be tested by incubating it with a cosmid and demonstrating linearization at the *cos* site. For improving packaging efficiency, we have found that the addition of 1 vol of crude terminase for each volume of sonicated extract (SE) gives about a 3-fold improvement in efficiency of ligated DNA. (It is important to realize that no benefit will be seen with control λ DNA, since the *cos* sites in such a λ preparation are unligated and do not require terminase for proper packaging.) The terminase is added after the SE, and then after an interval of 15 min the FTL (freeze − thaw lysate) extract is added in the usual amounts.

(vi)   Plate the test packaged material on a *supF*$^+$ strain (LE392 works well) and on a *supF*$^-$ strain (MC1061 is by far the best). The numbers in *Table 1* can be used as a guideline for reasonable expectations. Note that only the junction fragments bearing *supF* will be able to form plaques on MC1061, but all other genomic fragments must still be ligated and packaged, so that a large quantity of vector, as well as a large volume of packaging extracts, must be utilized to make a complete library.

(vii) Scale up the packaging reaction. By test platings, determine the maximum amount of packaged material which can be plated without losing efficiency. The library should then be plated at this density (or at 40 000 p.f.u./plate, whichever is less) on 150 mm plates, using MC1061 as the plating host. The library can then be screened, or amplified in standard fashion for repeated future screenings.

(viii) Screen the library by standard techniques. It is prudent to screen 20 000 or so clones from a new library with pBR322 as a probe in order to determine whether any of the plaques contain the plasmid sequences which frequently contaminate the *supF* gene preparation at low levels. If such clones are present, it will mean that all probes used for screening the library must be carefully gel purified in order to avoid mistaking these plasmid-bearing clones for true positives.

(ix)   Screening can be effectively done using several probes simultaneously. However, it is absolutely essential to include a positive control for each probe. Ideally, this should be a phage which contains that probe as part of its insert and which can be plated at low density so that the intensity to be expected of true positives can be directly visualized. A negative control consisting of the λ vector with a pBR322 insert should also be included with the screening so that these potential false positives can be recognized. The importance of these positive and negative controls cannot be overemphasized; these should also be used for the secondary and tertiary screenings as plaques are purified.

3.6 Evaluation of clones

Once clones are plaque purified, minilysate DNA can be prepared using any standard protocol. At this point the yields are slightly higher if one prepares the minilysates on LE392 rather than MC1061. Digestion of the DNA with *Eco*RI should reveal the size of the insert fragment. If more than one *Eco*RI fragment is present in a given clone, this is an indication that the two inserts have ligated into the vector during cloning. It is then appropriate to sandwich blot this minilysate DNA with the starting probe and with *supF* in order to determine if they are present on the same *Eco*RI fragment. If they are, this clone can be studied in the usual fashion. If they are not, then this is not a useful clone.

In order to further characterize jumping clones, it is useful to subclone the *Eco*RI fragment into a plasmid to allow easier manipulation. A convenient way to do this is to use the suppressor tRNA as a scorable marker so that the subclones can be readily identified (*Figure 8*). The digested minilysate DNA can be ligated into pBR322 which

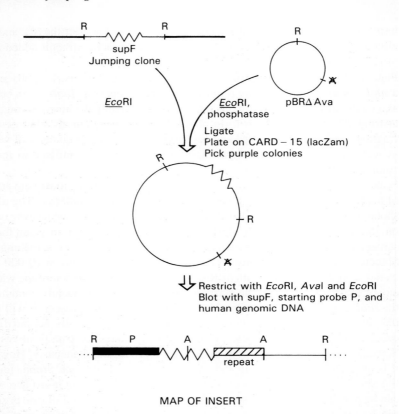

MAP OF INSERT

Figure 8. Scheme for rapidly subcloning inserts of jumping phage clones into pBRΔAva. R = *Eco*RI, A = *Ava*I.

has been *Eco*RI cut and phosphatased, and the resultant DNA transformed into a host strain which contains a *lacZ* amber mutation, such as CARD-15 (R.Dunn, personal communication). Plating on MacConkey's agar with ampicillin then reveals the desired colonies immediately, since they are capable of fermenting lactose and are purple whereas background colonies are pink.

To separate the starting point from the ending point of a jump, it is useful to take advantage of an *Ava*I site in the middle of the *supF* gene (24). This is particularly helpful if the *Ava*I site present in pBR322 has been destroyed, which we have done by *Ava*I digestion, Klenow fill-in and re-ligation. In this instance, one can perform digests of subclones with *Eco*RI, *Ava*I and the two together in order to separate the two halves of the jumping fragment. There are occasionally *Ava*I sites in the genomic sequences which makes it somewhat more difficult to construct the map, but blotting digests of the subclone with the starting probe, *supF* and human genomic DNA to identify the location of repeats usually allows the ready construction of a restriction map and the identification of single copy sequences which can be used to determine whether the jump is valid. Probing the blot with *supF* is particularly useful since it identifies the fragments which lie closest to the suppressor gene.

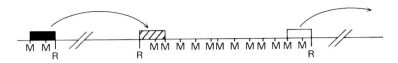

Figure 9. Sequential jumping in a general jumping library. In order for a particular sequence to appear in a library constructed as in *Figure 4*, that sequence must be contained on an *Eco*RI−*Mbo*I fragment. If used to re-screen the library, the jump fragment (cross-hatched) would in all probability lead to a jump back in the original direction (toward the black box). To continue to travel in the same direction, a probe (open box) near the far end of an *Eco*RI fragment would be much more successful. R = *Eco*RI, M = *Mbo*I.

Alternatively, if a *supF* has been used with additional rare restriction sites included (see *Figure 6*), separation of the two halves of the jump is easier and it is possible to prepare probes directly from phage minilysate DNA.

Once a part of the jump has been identified which is single copy, the fragment should be tested on a genomic blot of somatic cell hybrids to demonstrate that the jump remains on the correct chromosome. This is essential because of the risk of non-circular ligations during the jumping library construction.

3.7 Sequential jumping

Given the cloning strategy outlined in *Figure 4*, the fragments at the end of the jump are in general of length 300−5000 bp and abut an *Eco*RI site in the genome. A consideration of *Figure 9* indicates that if one uses this jumping fragment as a probe to go back into the same jumping library, it is quite likely that the clones derived will travel back in the direction of the original probe, rather than proceeding forward along the chromosome. There are two solutions to this problem. The first is to use a complementary library which uses a different restriction enzyme such as *Hind*III. As noted above, by alternating between *Eco*RI/*Mbo*I and *Hind*III/*Mbo*I libraries, it is more likely that one can continue to move in the same direction. Alternatively, it is reasonable to use the jump clone to obtain a larger genomic sequence from a standard λ library, that is to 'walk from the jump', prior to initiating the next jump. This will provide a larger number of probes to screen the jumping library, and will also provide a larger sequence to search for RFLPs. The latter is often an important part of the jumping procedure, since one wishes to check whether cross-overs between the original probe and the target have been traversed.

4. SPECIFIC JUMPING LIBRARIES

4.1 Principle

As diagrammed in *Figure 3* and discussed in Section 3.1, specific jumping libraries are constructed in such a way that they allow one to jump from a given rare restriction site to adjacent restriction sites for the same rare cutter. The basic principle, circularization of very large DNA molecules, is the same as for general jumping libraries, but in this instance the initial restriction digestion is a complete digest with a rare cutting enzyme, and there is no size selection of the resulting fragments. As pointed out in Section 3.1, an advantage of this sort of library is that far fewer clones are needed in order to represent all of the possible jumping fragments for a particular enzyme.

The obvious disadvantage is that one must be near one of those sites in order to use the library.

4.2 Protocol for library construction

The steps in library construction are quite similar to that for general jumping libraries outlined in detail in the previous section. Only the specific differences will be mentioned here. The description will be for the generation of a *Not*I jumping library, but can be readily generalized to jumping libraries for other rare restriction enzymes.

(i)   Since fewer clones are required in the final library, it is not necessary to start with as much DNA. In practice, we find it useful to use about 10 μg for the digestion, but usually 1 μg of this material is sufficient for eventual ligation and packaging. Carry out the initial restriction digestion with a considerable excess (10−20 U/μg) of enzyme, so that even sites which cut only partially in the genome will be represented in the jumping library. It is important to check the digestion of a portion of the DNA on a pulsed-field gel, as well as to confirm that DNA incubated in buffer without enzyme does not degrade under the conditions used.

(ii)   As before, recover the DNA from the agarose block either by melting or by electroelution.

(iii)   The suppressor tRNA gene around which the circles are generated must, of course, have the appropriate sticky ends. *SupF* genes with *Not*I and *Mlu*I ends have been generated; since *Mlu*I and *Bss*HII overhanging ends are identical, the *Mlu*I *supF* gene can also be used to generate a *Bss*HII jumping library. A potentially difficult issue is the concentration of *supF* to use in the ligation, since a wide range of genomic molecules are represented (16). When the DNA has been size selected, experiments have shown that the amount of *supF* added is not crucial so long as at least a 100-fold molar excess is present (F.Collins, unpublished data). Therefore, it is safest to add a sufficient amount of suppressor gene so that a 100-fold excess would be present even if all the molecules were 100 kb in length. Since, in general, the molecules are considerably larger than this, this guarantees a sufficient excess. An alternative is to do two circular ligations with two different concentrations of *supF*, one of which is appropriate for 100 kb length molecules and the other for 1000 kb length molecules. This approach was used by Poustka *et al.* (16) in their construction of a *Not*I jumping library; this may have been more important for the protocol they used than the one used here, because they phosphatased the marker gene.

(iv)   Choose the concentration for ligation so that it is enough for even the very large molecules in the mixture to stand an excellent chance of circularization. In practice, it is useful to work with concentrations of approximately 0.1 μg/ml, where even molecules of 2000 kb have a greater than 90% chance of circularization.

(v)   After precipitation, re-cut the genomic circles with *Eco*RI and clone the resulting fragments into λCh3AΔlac just as described in Section 3. A disadvantage of this approach is that jumping fragments which happen to reside on *Eco*RI fragments larger than 12 kb will not be clonable. A potential solution to this problem is to prepare an additional jumping library where the final digestion is with a different enzyme such as *Bam*HI. The resulting fragments can then be cloned into a zero-insert *Bam*HI cloning vector with amber mutations such as λCh30A.

(vi)   Check the validity of a *Not*I jumping library. This is somewhat easier, since valid clones will have both halves of the jump derived from the same *Not*I fragment. It is useful to identify 5−10 clones which contain single copy inserts, and then test the two halves of each jumping clone on a pulsed-field blot. This will give a good idea of the range of jump sizes represented in the library as well as the proportion which represent valid jumps.

(vii)   Subclone the separate halves of the jumping clone by digesting minilysate DNA with *Eco*RI plus *Not*I and ligating into a vector such as Bluescript which contains cloning sites for both these enzymes.

5. LINKING LIBRARIES

5.1 **Principle**

As diagrammed in *Figure 3*, specific jumping libraries are of the greatest utility when combined with linking libraries (4,5). Linking clones allow one to cross over a rare restriction site and then proceed to the next jump. Linking clones are simply fragments of DNA which contain the rare restriction site of interest. As such, linking libraries are easier to construct, since no manipulation of very large DNA molecules is necessary. While there are several possible ways to selectively clone genomic DNA fragments bearing a given rare restriction site, the method which we have found most efficient is diagrammed in *Figure 10*. This depends upon circular ligation of 15−20 kb size-selected partial *Mbo*I-digested genomic DNA around a suppressor gene. The circles are then cut with *Not*I; the majority will not cut but those which do can be selectively cloned into an amber-mutated phage vector capable of accepting DNA inserts of this size. The recently constructed λCh40A vector (25) is ideal for this purpose.

5.2 **Protocol for generation of *Not*I linking library**

(i)   Genomic DNA can be prepared by standard protocols and all manipulations can be done in liquid solution because there is no need to deal with extremely high molecular weight molecules. Carry out a partial *Mbo*I digestion and size selection over a sucrose gradient exactly as one normally would for generating a phage library (24).

(ii)   Dilute molecules in the desired size range (15−20 kb) to a concentration which will promote circularization (∼1 μg/ml) and add a 200-fold molar excess of purified *supF* gene with *Bam*HI ends. Carry out ligation as described in Section 3.

(iii)   Precipitate circularized DNA and then digest it with a large excess of *Not*I. It is a good idea to do a control digestion with a small amount of this material to which a plasmid with a known *Not*I site has been added, in order to ascertain that digestion is not being inhibited.

(iv)   Phenol-extract the digested DNA and re-precipitate. Then ligate this into the appropriate λ arms. Since most of the genomic DNA will still be circular, only a small amount of vector is needed in order to assure a 10:1 molar excess of vector arms to clonable insert.

(v)   Package the DNA and plate on a sup⁻ host such as MC1061. To facilitate subcloning the inserts, it is useful to have used a suppressor gene which has

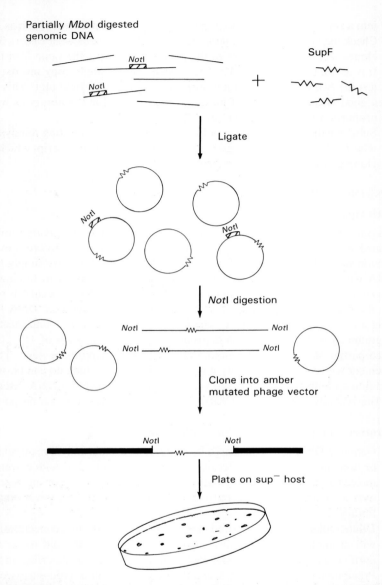

Figure 10. Strategy for making a *Not*I linking library. Only fragments bearing an internal *Not*I site can give rise to plaques in the library.

additional rare restriction sites internal to the *Bam*HI site, so that the two halves of the linking clone can be separately subcloned. Accordingly, we routinely use a suppressor which contains *Xho*I and *Sfi*I sites internal to the *Bam* site (see *Figure 6*).

(vi)   As with jumping clones, only about 10 000 *Not*I linking clones should be sufficient to represent the human genome.

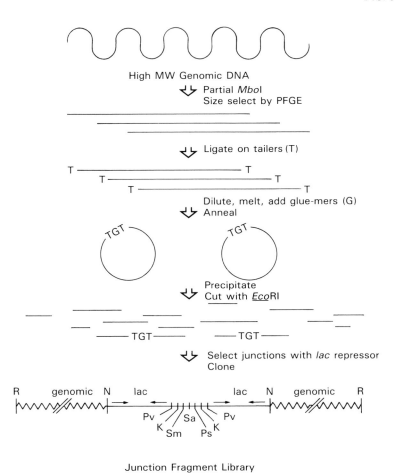

Figure 11. Scheme for constructing a jumping library using a *lac* operator – repressor system. Tailers (T) and glue-mer (G) are designed so that two inverted repeat *lac* operator sequences are generated after annealing, as well as a variety of restriction sites. N = *Not*I, Pv = *Pvu*I, K = *Kpn*, Sm = *Sma*I, Sa = *Sal*I, Ps = *Pst*I, R = *Eco*RI.

5.3 Use of sorted chromosome material

It is also possible to generate linking libraries from sorted chromosomal DNA, although the amount of material available (usually <1 μg) prohibits the partial digestion and size selection step. However, linking clones can also be generated from a complete digest, exactly as in *Figure 10* except that the initial step is a *Bam*HI or *Bgl*II complete digest. With this approach, we have constructed a library of 5000 clones from sorted chromosomes (M.R.Wallace, J.Gray and F.Collins, submitted). In this situation, the cloning vector must be able to accept smaller inserts. To this end, the λCh3AΔlac vector described above has been modified by replacing the *Eco*RI cloning site with a *Not*I site (K.Fry and J.Kim, unpublished).

6. POTENTIAL REFINEMENTS

While the protocols described up to this point have been successfully applied to the construction of general jumping libraries of 100 and 200 kb sizes (15,17) and specific jumping libraries which include jumps as large as 800 kb (16), there are significant technical challenges to the method which have slowed the application to even larger jump sizes. Among the major difficulties are:

(i)   the ligation volumes for circularization of general jumping libraries are often in the 25 – 100 ml range, which requires large amounts of DNA ligase;

(ii)   while only the junction fragments of the circles are desirable to clone, all other fragments must still be ligated to the λ vector and packaged before the *supF* selection can be applied; in practice this means that hundreds of microgrammes of vector and millilitre quantities of packaging extracts are required;

(iii)   especially when the size of genomic DNA is variable, as it is for rare cutter jumping libraries, it is difficult to choose conditions so that the majority of circles incorporate at least one *supF* gene;

(iv)   since the circularization is done by ligation, magnesium ions are required and the ligation must be allowed to proceed over long time periods to allow the ends of such large genomic molecules to find each other. Therefore, even tiny amounts of nuclease can be devastating.

Current efforts in jumping library technology are attempting to deal with these problems. A scheme of one developing protocol is shown in *Figure 11*. This basically deals with some of the difficulties noted above in two ways.

(i)   By adding linkers ('tailers') to the ends of the large genomic DNA molecules, circularization can be carried out by annealing rather than by ligation, which means that the reaction can proceed in the presence of EDTA to avoid problems with nuclease. The linkers are constructed in such a way that they are not self-complementary, but can be joined together by an additional oligomer, which we call a 'glue-mer'.

(ii)   The sequence of tailers and glue-mer is chosen such that two *lac* operator sequences are generated on annealing. This then allows the junction fragments to be selected by *lac* operator – repressor binding, which provides a physical selection rather than a biological one. The desired fragments can thus be selected prior to cloning, resulting in a substantial reduction in the amount of vector and packaging extracts required.

An additional improvement which is under investigation is the possibility of reducing the effective contour length of large genomic DNA fragments by complexing them with DNA-binding proteins. As is clear from the discussion in Section 3.2, the ability to shorten the contour length of DNA would significantly improve the possibilities of extending the protocol to larger and larger jump sizes. Simple approaches, such as altering salt concentration or using polyamines such as spermidine, both of which have a slight shortening effect on DNA contour length, do not result in a significant enough benefit to be practical. Living organisms, of course, have to deal with the same DNA packaging problem; it seems appropriate, therefore, to try to adapt the mechanisms already used by nature. Eukaryotic histones are one possibility, although reconstituting chromatin is not trivial, and there are reasons to fear that binding of histones near the ends of genomic fragments might block circularization. Another alternative which is being studied is to use the *Escherichia coli* HU protein (26), which also binds DNA and shortens its contour length by a factor of five. HU protein is relatively easy to prepare. It binds less tightly than histones, so that a dynamic equilibrium between bound and unbound DNA exists in solution, but with most of the DNA bound at any given moment. The looser binding, however, may be less troublesome for interference with circularization. While this is still in the experimental development stage, potentially this approach can allow 1000-kb molecules to be treated as easily as 200-kb molecules, which would considerably extend the range over which jumping approaches can be successfully applied. Until these new modifications are fully explored, it is impossible to predict what the ultimate limits of jumping technology may be.

7. ACKNOWLEDGEMENTS

I would like to thank Sherman Weissman, in whose laboratory the chromosome jumping concept was conceived, for his continued interest and input into these protocols. Several members of my laboratory have made major contributions to the refinements of this approach, particularly Jeffery Cole, Mitchell Drumm, Jane Fountain, Michael Iannuzzi, Julia Richards and Peggy Wallace. I would also like to thank Robert Dunn for providing the original *supF* gene, Frederick Blattner for supplying the λCh3AΔlac vector, Kirk Fry and Jungsuh Kim for creating the *Not*I version of this vector, Charles Cantor and Cassandra Smith for assistance in initially setting up PFGE, Hans Lehrach and Annemarie Frischauf for sharing protocols and results prior to publication, and Bernice Bishop for preparing the manuscript. I also acknowledge with gratitude support from the Howard Hughes Medical Institute, NIH Grant GM34960 and the Hereditary Disease Foundation.

8. REFERENCES

1. Ruddle,F. (1984) *Am. J. Hum. Genet.*, **36**, 944.
2. Orkin,S.H. (1986) *Cell*, **47**, 845.
3. Collins,F.S. and Weissman,S.M. (1984) *Proc. Natl. Acad. Sci. USA*, **81**, 6812.
4. Poustka,A. and Lehrach,H. (1986) *Trends Genet.*, **2**, 174.
5. Smith,C.L., Lawrance,S.K., Gillespie,G.A., Cantor,C.R., Weissman,S.M. and Collins,F.S. (1987) In *Methods in Enzymology*. Gottesman,M. (ed.), Academic Press, New York, Vol. 151, p. 461.
6. Schwartz,D.C. and Cantor,C.R. (1984) *Cell*, **37**, 67.
7. Carle,G.F. and Olson,M.V. (1984) *Nucleic Acids Res.*, **12**, 5647.
8. Carle,G.F., Frank,M. and Olson,M.V. (1986) *Science*, **232**, 65.
9. Chu,G., Vollrath,D. and Davis,R.W. (1986) *Science*, **234**, 1582.
10. Smith,C.L., Warburton,P.E., Gaal,A. and Cantor,C.R. (1986) In *Genetic Engineering*. Setlow,J.K. and Hollaender,A. (eds), Plenum Press, New York, Vol. 8, p. 45.
11. Smith,C.L. and Cantor,C.R. (1987) In *Methods in Enzymology*. Wu,R. (ed.), Academic Press, New York, Vol. 155, p. 449.
12. Burke,D.T., Carle,G.F. and Olson,M.V. (1987) *Science*, **236**, 806.
13. Botstein,D., White,R.L., Skolnick,M. and Davis,R.W. (1980) *Am. J. Hum. Genet.*, **32**, 314.
14. Donis-Keller,H. *et al.* (1987) *Cell*, **51**, 319.
15. Collins,F.S., Drumm,M.L., Cole,J.L., Lockwood,W.K., Vande Woude,G.F. and Iannuzzi,M.C. (1987) *Science*, **235**, 1046.
16. Poustka,A., Pohl,T.M., Barlow,D.P., Frischauf,A.-M. and Lehrach,H. (1987) *Nature*, **325**, 353.
17. Richards,J.E., Gilliam,T.C., Cole,J.L., Drumm,M.L., Wasmuth,J.J., Gusella,J.F. and Collins,F.S. (1988) *Proc. Natl. Acad. Sci. USA*, in press.

18. Shortman,K. (1972) *Annu. Rev. Biophys. Bioeng.*, **1**, 93.
19. Jacobson,H. and Stockmayer,W.H. (1950) *J. Chem. Phys.*, **18**, 1600.
20. Collins,F.S. (1986) In *Applications of DNA Probes*. Lerman,L. (ed.), Cold Spring Harbor Laboratory Press, Cold Spring Harbor, New York, p. 67.
21. Dunn,R.J., Belagaje,R., Brown,E.L. and Khorana,H.G. (1981) *J. Biol. Chem.*, **256**, 6109.
22. Hohn,B. and Murray,K. (1977) *Proc. Natl. Acad. Sci. USA*, **74**, 3259.
23. Rackwitz,H.R., Zehetner,G., Murialdo,H., Delius,H., Chai,J.H., Poustka,A., Frischauf,A. and Lehrach,H. (1985) *Gene*, **40**, 259.
24. Maniatis,T., Fritsch,E.F. and Sambrook,J. (1982) *Molecular Cloning—A Laboratory Manual*. Cold Spring Harbor Laboratory Press, Cold Spring Harbor, New York.
25. Dunn,I.S. and Blattner,F.R. (1987) *Nucleic Acids Res.*, **15**, 2677.
26. Broyles,S.S. and Pettijohn,D.E. (1986) *J. Mol. Biol.*, **187**, 47.

CHAPTER 5

Detection of single base changes in DNA: ribonuclease cleavage and denaturing gradient gel electrophoresis

RICHARD M.MYERS, VAL C.SHEFFIELD and DAVID R.COX

1. INTRODUCTION

Techniques that allow the detection of single base changes in genomic DNA have had a major impact on our understanding of human genetic diseases, both by identifying specific mutations that result in disease (1−6), and by identifying DNA polymorphisms that are used as genetic markers in linkage studies (7−11). In addition, these experimental techniques have made it possible to develop accurate pre- and post-natal diagnostic tests for several human genetic diseases. In cases where the gene involved in a disease is known and the DNA sequence at, and immediately adjacent to, a specific mutation is available, the mutation can be detected in genomic DNA or in RNA by using labelled oligonucleotides as hybridization probes to blotted test samples. In cases where the DNA sequence in the region of a specific mutation is not known, base changes have been identified as restriction fragment length polymorphisms (RFLPs; 7). RFLPs are detected by determining the presence or absence of a restriction enzyme cleavage site within a DNA fragment in genomic DNA samples, generally by hybridizing a labelled DNA probe to restriction-digested genomic DNA that has been size fractionated in an agarose gel and transferred to a filter membrane. This method has been extremely powerful in identifying both mutations and neutral polymorphisms in studies of humans and other organisms. However, because the chances that a base change will alter a restriction site are low, the RFLP method fails to detect a large fraction of mutations and polymorphisms. For example, many of the single base mutations in the human β-globin gene that are known to cause thalassaemia do not alter a restriction enzyme cleavage site, and so cannot be directly detected by the RFLP method (12). In addition, it appears that some regions of the mammalian genome are sparse with respect to polymorphisms, making it extremely difficult to find RFLPs in these regions even when a large number of different restriction enzymes are used (13,14). For these reasons, new strategies for detecting single base changes in genomic DNA samples would be valuable.

In collaboration with Dr Tom Maniatis and Dr Leonard Lerman, two new methods for detecting single base changes in genomic DNA samples were developed in an effort to supplement the RFLP approach (15−20). These methods, called RNase cleavage and denaturing gradient gel electrophoresis (DGGE), each allow the detection of at

95

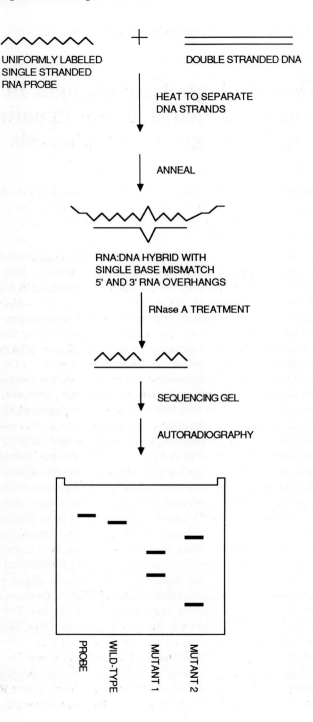

least 50% of all possible base changes within a segment of genomic DNA up to 1000 bp in length.

The focus of this chapter is to discuss these methods in detail, describe their advantages and disadvantages, provide a trouble-shooting guide, and present some new modifications of the published procedures that make them easier to use. In addition, we describe how these methods can be adapted to examine DNA fragments generated by the amplification of specific sequences in a small amount of genomic DNA by the polymerase chain reaction (PCR; 21−23).

2. GENERAL DESCRIPTION OF THE METHODS

2.1 RNase cleavage

Single base changes in both cloned and genomic DNA fragments have been detected by cleavage of mismatches in RNA:DNA duplexes with RNase A (15,16). The general approach involves the following steps, which are depicted in *Figure 1*.

(i)   Synthesis of a uniformly labelled single-stranded RNA probe by *in vitro* run-off transcription of a cloned DNA fragment.

(ii)   Hybridization of the probe to its complementary sequences in cloned, genomic or PCR-amplified DNA samples, which results in an RNA:DNA duplex containing a single base mismatch if a base change is present in the DNA covered by the probe.

(iii) Treatment of the duplex with RNase A, resulting in cleavage of the RNA strand at many (though not all) mismatch positions.

(iv)   Analysis of the labelled RNA products by denaturing gel electrophoresis followed by autoradiography. Mismatches efficiently cleaved by the procedure result in the generation of two labelled RNA fragments on the autoradiogram, the sizes of which indicate the position of the base change relative to the ends of the DNA fragment. A single labelled RNA fragment, whose length is equal to the full-length DNA fragment, is observed if no base change is present in the DNA sample, or in cases where the base change does not result in cleavage of the mismatch in the RNA:DNA duplex.

Several examples of the data obtained in RNase cleavage reactions are shown in *Figure 2*.

Figure 1. The RNase cleavage procedure. A uniformly labelled single-stranded RNA probe is synthesized as a run-off transcript from a cloned DNA template. The probe is mixed with double-stranded test DNA, which can be cloned, genomic or PCR-amplified DNA fragment, and the mixture is heated to separate the DNA strands. An annealing reaction is performed to allow duplexes to form between the labelled RNA probe and its complementary strand in the test DNA fragment. If a base change is present in the test DNA sample, a single base mismatch will be present in the RNA:DNA hybrid. The annealed sample is then treated with RNase A, which cleaves at many mismatches in the RNA strand. Note also that the RNase A removes the RNA 'overhangs' at the 5' and 3' ends of the duplex. The duplex is then treated with denaturants to separate the strands, and run on a denaturing ('sequencing') gel to separate the RNA fragments by size. The gel is dried and exposed to X-ray film. If the test DNA had no mutation, a single band on the autoradiogram, corresponding in length to the full-length protected test fragment, is observed. If a mutation was present in the test DNA, two bands will be observed on the autoradiogram.

Detection of single base changes in DNA

Approximately 30−40% of all the possible mismatches in an RNA:DNA duplex are cleaved efficiently by this procedure. However, by testing a DNA fragment with each of its two corresponding labelled RNA strands in separate cleavage reactions, about 60−70% of all possible base changes in a DNA fragment can be detected since the complementary mismatches of most of the poorly-cleaved mismatches are cleaved efficiently (see discussion in 15,16).

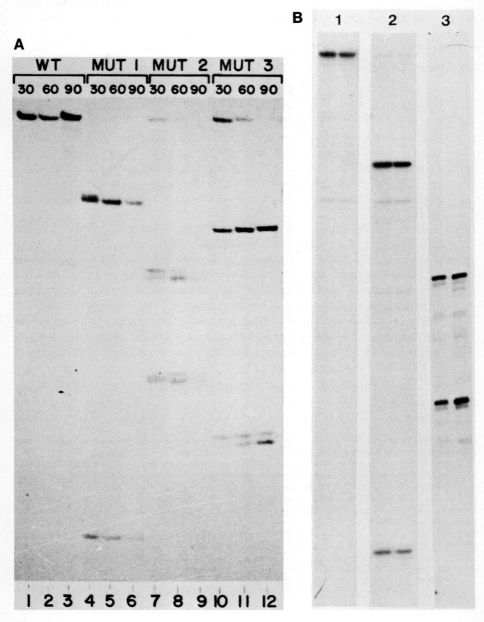

98

In our hands, the optimal size of the RNA probe, and thus the size of the region being screened for base changes, is between 100 and 1000 nucleotides. With probes in this size range, the full-length and cleaved RNA products can be readily visualized following electrophoresis in denaturing (urea) polyacrylamide gels of the type used for DNA sequencing (24), and the signal-to-noise ratio is high enough to give unambiguous results. Although it is easy to generate single-stranded labelled RNA probes much longer than 1000 nucleotides, we have been unable to achieve adequately clean results in the RNase cleavage reaction with long probes. Under conditions of RNase treatment that are needed for cleavage at mismatched positions in RNA:DNA hybrids, enough random cleavage of RNA strands at perfectly-paired positions occurs to present background problems when probes longer than 1000 nucleotides are used. In addition, analysis of RNA products longer than 1000 nucleotides requires the use of denaturing agarose gel electrophoresis. Occasionally we have had difficulty achieving complete denaturation of the RNA:DNA hybrids on such gels, and the results are confusing because completely cleaved mismatches appear as full-length fragments on the gel if the duplexes are not completely denatured. We have also attempted to increase the efficiency of screening for single base changes by RNase cleavage by using more than one probe in a single annealing/cleavage reaction. Although this approach has been successful, in many cases the results obtained with multiple probes have been too complicated or too messy to interpret. For these reasons, we recommend that the RNase cleavage procedure be used to examine single regions of DNA defined by RNA probes 100−1000 nucleotides in length.

An earlier report by Freeman and Huang (25) indicated the ribonuclease can cleave at presumed single base mismatches in RNA:RNA duplexes. Indeed, M.Perucho and his colleagues (26) found that confirmed single base mutations in the H-*ras* gene can be detected by cleaving mismatches in duplexes made between H-*ras* mRNA and an RNA probe of the opposite sense. In addition, we have found that naturally-occurring β-globin mutations and synthetically-generated β-interferon promoter mutations that are detected by RNA:DNA cleavage can also be detected by cleavage of RNA:RNA

Figure 2. RNase cleavage reactions. (**A**) RNase cleavage reactions with cloned test DNA samples. A uniformly-labelled RNA probe was annealed to a small amount of cloned DNA, treated with RNase A and examined by denaturing gel electrophoresis. The autoradiogram of the gel shows the results of cleavage of a perfectly-paired RNA:DNA duplex ('WT' lanes) and of RNA:DNA duplexes containing a single base mismatch ('MUT1, MUT2 and MUT3' lanes). Three time points of RNase cleavage are shown for each DNA tested, 30, 60 and 90 min of cleavage. In the 'WT' lanes, no cleavage of the duplex occurs, and the labelled RNA product appears at the position in the gel corresponding to full-length protected RNA probe. In each of the three mutants tested, cleavage at the single base mismatch occurs, resulting in two additional RNA products whose lengths equal that predicted for proper cleavage. In MUT2 and MUT3, the cleavage is incomplete at the shorter reaction times, resulting in a fraction of full-length probe. With continued incubation, these partially-cleaved duplexes are cleaved to completion (see 90 min time point). (Reprinted with permission from ref. 15.) (**B**) RNase cleavage reactions with genomic test DNA samples. A uniformly-labelled RNA probe comprised of the first exon, first intron and part of the second exon sequences in the human β-globin gene was annealed to about 5 μg of total genomic DNA from three individuals. After allowing RNA:DNA duplexes to form, the mixture was treated with RNase A under conditions described in the text. The resulting samples were denatured and loaded onto a 8% polyacrylamide/8 M urea gel, electrophoresed and the gel was exposed to X-ray film. Each set of two lanes contains duplicate RNase cleavage reactions: (**1**) genomic DNA from an individual with normal β-globin genes, (**2**) genomic DNA from an individual with β-thalassaemia homozygous for a mutation at codon 26 (HbE) and (**3**) genomic DNA from an individual with β-thalassaemia homozygous for a mutation at position 110 in the intron.

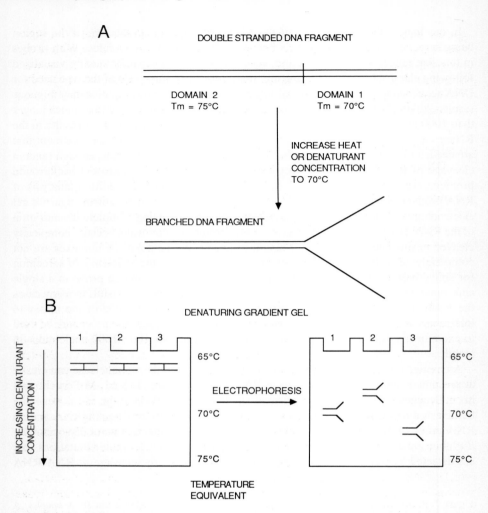

Figure 3. Melting behaviour of DNA in solution and in denaturing gradient gels. (**A**) The top drawing represents the double-stranded form of a DNA fragment at physiological temperatures. This DNA fragment, which is 100–1000 bp in length, melts in two domains, with Tms of 70 and 75°C. As the temperature or denaturant concentration of a solution of the DNA is gradually raised to 70°C, the first domain melts, resulting in the branched molecule shown in the bottom drawing. A further increase in denaturant (to 75°C) results in complete strand separation of the fragment. (**B**) The use of DGGE to separate DNA fragments differing by a single base change. The gel on the left shows three double-stranded DNA fragments at the beginning of an electrophoretic run entering a polyacrylamide gel containing a linearly-increasing gradient of denaturants equivalent to 65°C at the top and 75°C at the bottom. Note that the DNA fragments are completely double-stranded. The gel on the right shows the DNA fragments at their final positions in the gel after electrophoresis. The first domain has melted in each of the fragments, so they appear as branched molecules. The DNA fragment in **lane 1** corresponds to a 'wild-type' fragment with a Tm of 70°C for its first domain. **Lane 2** shows a mutant DNA fragment that has a lower Tm for its first domain, so it begins to melt and slow down in the gel at lower denaturant concentrations. The mutant DNA fragment in **lane 3** melts at a higher temperature, so it travels further into the gel before slowing down. The mobility retardation resulting from the partial melting of DNA fragments in DGGE causes the DNA fragments to focus sharply in the gel, allowing very fine resolution of bands.

duplexes. In these experiments, a uniformly labelled single-stranded anti-sense RNA probe was made from a cloned DNA template as usual, and the probe was annealed to its complementary mRNA, the mixture treated with RNase and the products were examined by denaturing polyacrylamide gel electrophoresis. Although the detailed experimental methods below are described for testing DNA samples, we have found that identical conditions can be used for testing RNA.

2.2 Denaturing gradient gel electrophoresis

Two DNA fragments differing by changes as simple as a single base substitution, deletion, insertion or mismatch can be physically separated from each other by a procedure called DGGE (15,17−20,27,28). This method involves the electrophoresis of double-stranded DNA fragments through a standard acrylamide gel containing a linear gradient of DNA denaturants, such as formamide and urea or temperature. The theoretical basis of the gel system is described in detail below; stated simply, the separation in the gel is due to the fact that two very similar DNA fragments often have different melting temperatures.

2.2.2 *Theory of DGGE*

As temperature or denaturant concentration is gradually raised, a double-stranded DNA fragment in solution goes through a distinct pattern of melting behaviour. Blocks of sequence within the fragment, called melting domains, melt cooperatively at discrete temperatures, called a 'Tm' for a given domain (*Figure 3A*). Melting domains are between 25 to several hundred base pairs in length. Two adjacent domains may differ in Tm by several degrees and have fairly sharp boundaries. DNA fragments 100−1000 bp in length generally have two to five melting domains. The melting domains of restriction fragments from most naturally-occurring sources are between 65 and 80°C.

It is common knowledge that the base composition of a DNA fragment affects its Tm; many procedures for estimating optimal annealing and washing conditions for nucleic acid hybridization reactions rely on knowledge of the GC content of the reactants. What is less well appreciated is the large contribution that stacking interactions make in determining the thermodynamic stability of the double helix. Stacking interactions between adjacent bases on the same strand of a DNA fragment stabilize the DNA in the helical form, contributing more energy to stabilization than that contributed by the hydrogen bonds used in base pairing of opposite strands in physiological solutions (29). The order of the bases on a strand determines the degree of stacking, and hence the degree of stability, that will occur locally. Changes in base sequence as small as a single base substitution can alter the stacking significantly enough to change the Tm by over 1°C. Because the melting of a domain is extremely cooperative, a single base change at any position in a domain will alter its Tm to some extent.

Leonard Lerman and Stuart Fischer devised a means for exploiting these sequence-dependent melting differences so that two very similar DNA fragments can be separated from one another electrophoretically (27,28). Their system is based on the fact that base sequence affects the Tm of a domain, and that the conformation of a DNA fragment affects its migration in a gel matrix when driven by an electric field. The principle is to electrophorese a DNA molecule through a gel of fixed acrylamide concentration that contains a linearly-increasing gradient of DNA denaturants, usually formamide and urea, from top to bottom (*Figure 3B*). The electrophoresis apparatus is set at a relatively high temperature, usually 60°C, that is near but below the Tm of the lowest temperature melting domain in the DNA fragment. The DNA fragment enters the gel as a double-stranded molecule, and migrates at a linear rate dependent on its molecular weight. As the DNA fragment enters the position in the gel where the denaturant concentration and bath temperature equal the temperature of the Tm of the first melting domain, the fragment suddenly becomes branched, with a double-stranded portion and a single-stranded portion. This branched structure becomes entangled in the pores of the gel matrix, and consequently its mobility is retarded. The degree of mobility retardation is roughly a function of the lengths of the melted and unmelted regions; the longer the first melting domain, the greater the degree of mobility retardation. The position in the gel at which the DNA fragment begins to slow down is equivalent to the Tm of the lowest temperature melting domain of the fragment. If the denaturant concentrations of the gel are chosen correctly, two DNA fragments differing by a single base substitution that have different Tms will begin slowing down at different positions in the gel, and will be separated from each other at the end of the run.

Originally it was thought that the degree of retardation would be such that DNA fragments with base changes in only the lowest temperature melting domain would be separated from the wild-type fragment on denaturing gradient gels. However, analysis of a large number of DNA fragments containing single base changes, as well as subsequent theoretical formulations, indicated that, for most fragments, base changes in all but the highest temperature melting domain are accessible to analysis by DGGE (18,19). Base changes in the last domain to melt are generally not detectable by the gel system because as this domain melts, the DNA fragment undergoes complete strand separation, resulting in the loss of sequence-dependent mobility. A key principle to remember in DGGE is that a branched DNA fragment is required for separation of wild-type and mutant DNA fragments, where the mutations fall in the melted regions of the fragments. The gel system has been improved somewhat by attaching a high-temperature melting domain, which is designated a 'GC-clamp', next to DNA fragments with domains of lower temperature, thus rendering the entire attached fragment accessible to analysis (18,19). This improvement has been demonstrated with cloned test DNA samples by cloning the sequence in question into a plasmid vector containing a GC-clamp. Recently, we have used PCR (see Section 2.3) to attach without a requirement for a cloning step.

A second improvement in DGGE is the use of heteroduplexes between wild-type and mutant DNA fragments. Earlier experiments involved the comparison of mobilities of wild-type and mutant homoduplex fragments in side-by-side lanes in DGGE. By forming heteroduplexes between a wild-type fragment and a mutant DNA fragment, which results in a DNA fragment containing a single base mismatch, the resolution and fraction of mutations detected by the gel system can be greatly increased (15,17,20). The reason for increased resolution is that mismatches in a DNA fragment tend to cause enough destacking of bases that a large decrease in stability occurs, thus making it easier to melt the DNA fragment. In some cases, we have seen destabilization by as much as 6°C due to a single base mismatch (15). In fact, analysis of a large number of heteroduplexes as well as theoretical calculations indicate that 100% of all possible changes in the lower temperature melting domains of a DNA fragment have a detectable shift in mobility in DGGE when in the mismatched form (R.M.Myers, T.Maniatis and L.S.Lerman, unpublished). On average, over half of the length of DNA fragments in the size range of 100 – 1000 bp fall into lower temperature melting domains, so over 50% of all possible base changes in these fragments can be detected by DGGE if heteroduplexes are used. Furthermore, with cloned DNA fragments adjacent to a GC-clamp, all possible single base changes in the attached fragment are detectable (18,19).

Heteroduplex analysis also makes it possible to see altered mobilities of mutant DNA fragments that would normally not be detectable by the gel system. In many cases, this ability to detect new mutations is due to a large alteration in the domain structure of the DNA fragment caused by the single base mismatch. This sort of large-scale change in melting behaviour can switch the order of domain melting so that a highest-temperature domain can become the first domain to melt (R.M.Myers, T.Maniatis and L.S.Lerman, unpublished). Thus, some base changes in the highest-temperature domain can become accessible to analysis by DGGE when heteroduplexes are examined instead of homoduplexes.

2.2.2 *DGGE in practice*

The schemes that are described here have all been designed to take advantage of the increased resolution and efficiency of detection of heteroduplex analysis. A general strategy is depicted in *Figure 4*. Test DNA samples (cloned, genomic or PCR-amplified) are annealed to a radioactively labelled, single-stranded DNA probe, generating a heteroduplex with a single base mismatch in the cases where a base change is present in the test sample. The test samples are then run in a denaturing gradient gel next to a lane containing the wild-type homoduplex. Following electrophoresis, the gel is examined by autoradiography. Thus, no gel blotting is required. In addition to using single-stranded DNA probes in DGGE, it is possible to use asymmetrically labelled double-stranded DNA probes (see Section 6.3), and labelled single-stranded RNA probes to test DNA in RNA:DNA hybrids (Section 6.3; S.Wolf, personal communication) or to test RNA in RNA:RNA hybrids (31). Several of these approaches are discussed in Section 6.

Similar to the RNase cleavage procedure, the DGGE system can be used to examine double-stranded nucleic acid fragments (DNA, RNA:DNA and RNA) of about 100 – 1000 bp. The upper limit of size is partially determined by the fact that the gel matrix is polyacrylamide. This matrix slows the mobility of larger DNA fragments dramatically so that very long electrophoretic run times are required to separate fragments larger than 1000 bp. A more serious problem is due to the melting properties of larger DNA fragments; the larger the fragment, the more melting domains it will have, and thus severe mobility retardation due to melting of multiple domains will often occur early in the gel run, rendering only a small portion of the fragment accessible to base change detection. For these reasons, the limit of size of fragments that we routinely examine

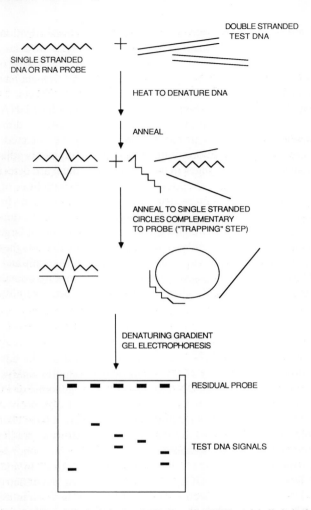

Figure 4. A scheme for detecting single base substitutions by DGGE. A labelled single-stranded DNA or RNA probe is mixed with double-stranded cloned, genomic or PCR-amplified DNA, and the mixture is heated to separate the DNA strands. An annealing reaction is performed, resulting in hybrid duplexes between the probe and its complementary strand in the test DNA sample. If a mutation is present in the test DNA sample, a single base mismatch will be present in the probe:test strand duplex. After the annealing reaction, an excess of single-stranded circular DNA complementary to the probe is added to the mixture to bind any residual probe. The mixture is then subjected to electrophoresis in a denaturing gradient gel. The gel is dried and exposed to X-ray film. Mismatches in mutant DNA samples that alter the melting behaviour of the fragment cause the fragments to shift upward in the gel due to their decreased thermal stability. Note that this procedure takes advantage of the increased resolution on DGGE of mismatched molecules.

is 1000 bp. Because on average, greater than 50% of the length of most DNA fragments falls in the lower temperature melting domains, examination of a 1000-bp fragment by DGGE tests approximately 500 of the nucleotides in the fragment for base changes. It is possible to examine more than one DNA fragment in a single lane by DGGE. A convenient way to maximize the information gained from an analysis is to use a DNA probe 1000–2000 bp in length, anneal it to genomic DNA and digest the duplex with restriction enzyme that creates two or three fragments of optimal size (i.e. 500–700 bp per fragment).

Several examples of data obtained with DGGE are shown in *Figure 5*.

A comparison of the use of RNase cleavage, DGGE and RFLP analysis to detect single base changes in DNA is presented in *Table 1*.

2.3 **The polymerase chain reaction (PCR)**

Both the RNase cleavage and DGGE methods can be used to examine genomic DNA samples directly, without cloning each test DNA sample. By using uniformly labelled probes of fairly high specific radioactivity (~200–400 Ci/mmol in one of the nucleotides), it is possible to obtain results with either method starting with 5–10 µg of human genomic DNA and using 24 h autoradiographic exposures. With genomic DNA from less complex organisms the sensitivities are correspondingly increased. Although the results obtained as such are usually of high quality, there are several problems with directly using total genomic DNA. First, the signal-to-noise ratio is low enough that background sometimes becomes a problem, particularly with the RNase cleavage procedure. Second, the probes must be uniformly labelled with ^{32}P to such a degree that they must be used within a day or two to avoid background from radioactive decay. Third, the amount of genomic DNA required for the analysis is sometimes prohibitive, especially when multiple tests need to be run. Any test is better if smaller amounts of genomic DNA could be used.

In 1985, a remarkable new procedure called the PCR was reported (21–23); this method allows specific regions of DNA within a genomic DNA sample to be amplified by as much as a million-fold. The procedure is simple in concept and execution (*Figure 6A*). First, two deoxyoligonucleotides, 20–30 nucleotides in length, and composed of the nucleotide sequences at the two ends of a DNA segment, are synthesized. The polarity of the two oligonucleotides is chosen so that they face each other in a 5' to 3' direction (see *Figure 6A*). Next, excess amounts of these two oligonucleotides are mixed with genomic DNA, and the mixture is heated to separate the genomic DNA into single strands. The temperature is then lowered so that the oligonucleotides will anneal with homologous sequences in genomic DNA. Following this step, an extension reaction with deoxyribonucleotide triphosphates and a DNA polymerase is performed. This sequence of denaturation, annealing and extension reactions is repeated 20–30 times. After two cycles, some of the products of the reaction correspond to a DNA fragment exactly the size expected for a complete end-to-end fragment demarcated by the two oligonucleotides (*Figure 6A*). This product serves as template for subsequent reactions, and is identical to most of the product at the end of the multiple cycles. Because the products of the reaction serve as templates for subsequent reactions, the procedure is indeed a chain reaction; the amount of product grows geometrically. Thus, with 100% efficiency at each step, 20 cycles will result in 2^{20} ($>10^6$) molecules. In practice, the efficiency is usually between 20 and 50%, so it is quite feasible to obtain over a million-fold amplification of specific sequences if enough cycles are run.

If oligonucleotides 20 nucleotides or longer are used for the PCR reaction to amplify sequences in vertebrate genomic DNA, the amplification is specific enough to produce

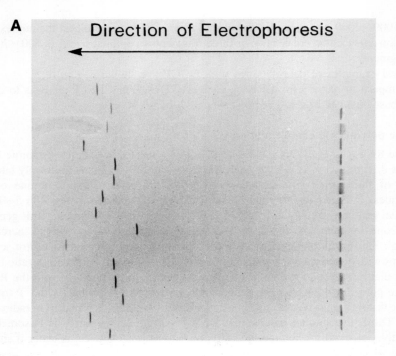

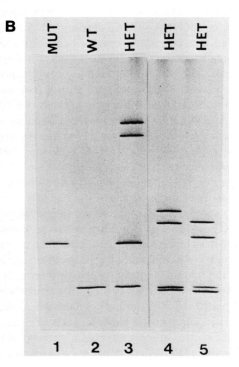

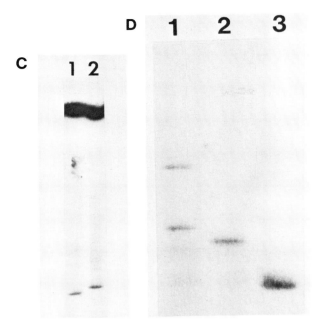

Figure 5. Denaturing gradient gel electrophoresis. (**A**) Analysis of cloned homoduplex DNA fragments by DGGE. This photograph is the negative image of an ethidium-stained denaturing gradient gel in which DNA fragments 450 bp in length and differing by single base substitutions were analysed. The lane on the left contains a wild-type reference sequence of the DNA fragment, and the remaining lanes contain individual cloned mutant DNA fragments. (Reprinted with permission from ref. 30.) (**B**) Analysis of cloned heteroduplex DNA fragments by DGGE. A negative image of an ethidium-stained denaturing gradient gel containing a mutant homoduplex (**lane 1**), a wild-type homoduplex (**lane 2**) and a mixture of homoduplex and heteroduplex fragments formed by mixing the wild-type and mutant homoduplexes in equimolar amounts, denaturing and reannealing to allow the re-assortment of the strands to the four expected species (**lane 3**). The two heteroduplex species in **lane 3**, each of which carries a single base mismatch, are greatly retarded in the gel due to the large destabilization in melting temperature caused by the mismatches. Additional homoduplex and heteroduplex mixtures, generated with the same wild-type DNA fragment and two different mutant DNA fragments, are shown in **lanes 4** and **5**. The importance of using heteroduplexes to increase resolution in DGGE is most apparent in **lanes 4** and **5**, where the wild-type and mutant homoduplexes are barely separated, and the heteroduplexes separate with much higher resolution from the wild-type homoduplex. (**C**) Analysis of total genomic human DNA by DGGE. This panel contains an autoradiogram of a denaturing gradient gel in which DNA samples from two individuals was examined. A single-stranded DNA probe composed of the sense strand of the human β-globin gene from nucleotides 132 to 404 was annealed to 5 μg of total genomic DNA from an individual with normal β-globin genes (**lane 1**) and an individual with β-thalassaemia homozygous for a G-to-A mutation at codon 39 (**lane 2**), the residual probe was 'trapped' with a single-stranded circular plasmid DNA complementary to the probe, and the mixture was loaded onto a 30−60% denaturant, 14% acrylamide gel. After electrophoresis at 150 V for 10.5 h at 60°C, the gel was dried and subjected to autoradiography for 20 h at −70°C with an intensifier screen. The 272-bp globin fragments are seen at the lower part of the gel, and the residual excess trapped probe is seen at the top part of the gel near the origin of electrophoresis. (Reprinted with permission from ref. 15.) (**D**) Analysis of PCR-amplified total human genomic DNA by DGGE. This panel contains an autoradiogram of a 30−60% denaturant, 14% acrylamide gel in which DNA samples from two individuals with β-thalassaemia (**lanes 1** and **2**) and one individual with normal β-globin genes (**lane 3**) were analysed. Approximately 300 ng of total genomic DNA from these individuals was amplified by the PCR reaction as described in the text, and 1/100th of the amplified DNA was annealed to a single-stranded labelled probe similar to that described in (**C**). One tenth of the annealed sample was subjected to gel electrophoresis, and the gel was dried and exposed to X-ray film with an intensifier screen for 6 h. Thus, with PCR amplification, the equivalent of 2−3 ng total genomic DNA results in signals that can be seen in just a few hours of autoradiography. The genomic DNAs were from individuals with globin genes heterozygous for mutations at intron 1, positions 1 and 6 (**lane 1**), homozygous for a mutation at intron 1, position 110 (**lane 2**) and homozygous for the normal β-globin sequence (**lane 3**).

107

Table 1. A comparison of the properties of RFLP analysis, RNase cleavage and denaturing gradient gel electrophoresis.

	No. of bp screened per lane	Gel	Detection method	Probe	Size range of test fragment (bp)	Quantity DNA needed/lane	Preliminary steps	Other considerations
RFLP	10–20	Agarose	Blot, autoradiography	(i) Nick-translated (ii) Random-primed (both double-stranded)	500–10 000	5–10 µg	(i) Clone probe fragment into vector (ii) Gel purify probe fragment	(i) Small fraction of base changes are detected (ii) Single base changes that produce small DNA fragments are not detected
RNase cleavage	250–500	Urea/acrylamide (sequencing gel)	Direct autoradiography	Uniformly labelled single-stranded RNA	100–1000	3–5 µg with PCR <10 ng	(i) Clone probe fragment into RNA-pol vector (ii) Simple restriction mapping (iii) For PCR: sequence ends of probe fragment	(i) Cleavage at perfectly paired positions can cause background (ii) Incomplete cleavage can make heterozygote detection difficult
DGGE	250–500	Acrylamide with urea/formamide gradient	Direct autoradiography	(i) Uniformly labelled single-stranded DNA or RNA (ii) End-labelled single-stranded DNA	100–1000	3–5 µg with PCR: <10 ng	(i) Clone probe fragment into vector (ii) Simple restriction mapping (iii) Perpendicular gel (iv) Travel schedule gel (v) For PCR: sequence ends of probe fragment	Considerable initial set-up investment

a single amplified fragment of DNA. However, with some amplification reactions, other unexplained DNA fragments are generated. Because these fragments occasionally appear to interfere with subsequent analysis of the specific DNA fragment by RNase cleavage, DGGE or DNA sequencing, a second round of amplification with a second set of oligonucleotides is recommended. This method, called the 'nested oligo' procedure (22), involves using a small fraction of a first PCR reaction, in which a large DNA fragment is produced, as the template in a second round of PCR with two new oligonucleotides that hybridize at 'internal' positions within the first amplified fragment (*Figure 6B*). When this approach is used, a large amount of a single, specific, shorter DNA fragment is produced, which is suitable for all subsequent analysis.

PCR routinely results in several microgrammes of a specific DNA fragment from less than 1 μg of starting total genomic vertebrate DNA. The amplification works well for fragments up to 2000 bp (possibly longer), and several fragments can be amplified simultaneously in the same reaction. PCR can be used to amplify specific fragments for many applications in diagnostic and cloning procedures. Levinson and Gitschier have used PCR-amplified genomic DNA and RNase cleavage to detect single base changes in the human factor VIII gene that result in the X-linked disease Haemophilia A (32). In the experimental methods sections below, we describe how PCR can be combined with both RNase cleavage and DGGE to detect single base changes in genomic DNA samples. For a more thorough coverage of the method, we refer the reader to the original publications (21−23) and subsequent publications (33,34). The method is also discussed in detail in Chapter 6.

3. PRELIMINARY PREPARATIONS

Several preliminary steps must be taken before the PCR, RNase cleavage or DGGE procedures can be used. These are listed below.

(i)   The DNA fragment that is to be tested for mutations or polymorphisms must be cloned into a plasmid vector that allows the synthesis of some type of probe, generally a single-stranded RNA or DNA probe. Several of these vectors are described in Sections 5.1 and 6.3.

(ii)   A restriction map of the cloned DNA insert is often helpful and sometimes necessary. The locations of restriction sites, preferably ones that appear only once or twice in the entire plasmid, are determined by standard restriction mapping procedures (24). For both single-stranded RNA and DNA probes, it is necessary to have a single-cutter restriction enzyme site at the end of the test DNA insert that is distal to the RNA polymerase binding site (for RNA probes; see Section 5.1) or the oligonucleotide binding site (for DNA probes; see Section 6.3 and *Figure 10*). It is also sometimes useful to use restriction sites within a test DNA fragment to cut the DNA hybrid resulting from annealing a uniformly labelled single-stranded DNA probe to a test DNA sample into two or three optimal size fragments for DGGE.

(iii)   If PCR is used, it is necessary to determine the nucleotide sequence of the two ends of a test DNA insert. Generally, only 40−50 bp of sequence information needs to be obtained to find a suitable, unambiguous 20−25 bp stretch for a first set of oligonucleotides. If the nested oligo procedure is to be used, additional sequence at any position internal to the first set of oligonucleotides will be needed.

4. THE POLYMERASE CHAIN REACTION

The original PCR method used the large fragment of *Escherichia coli* DNA polymerase as the DNA-synthesizing enzyme (21). More recently, it was found that the heat-resistant DNA polymerase isolated from *Thermus aquaticus* is more suitable for PCR for several reasons. First, this enzyme survives the many cycles of heating, annealing and polymerization, so that additional aliquots of enzyme do not have to be added after each cycle, resulting in a great saving in labour and time. Second, because the extension reactions are carried out at relatively high temperature, secondary structure in the DNA template that often interferes with *in vitro* elongation of DNA polymerases is abolished.

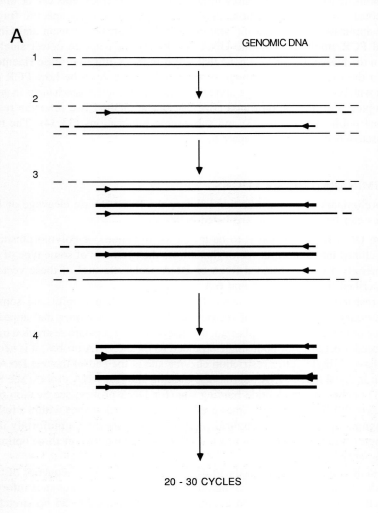

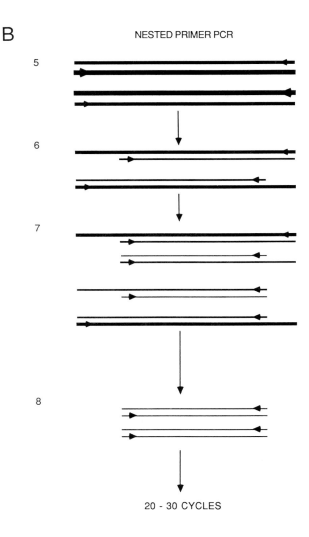

Figure 6. The polymerase chain reaction (PCR). (**A**) Genomic DNA (**1**) is denatured and annealed with an excess of two oligonucleotides, depicted as small arrows in (**2**). The annealed oligonucleotides are extended with dNTPs and a heat-resistant DNA polymerase (**2**). The DNA is denatured, annealed to the oligonucleotides and extended a second cycle to produce the products shown in (**3**). The next cycle of heating, annealing and extending produces the products shown in (**4**) as well as those shown in (**3**). Following an additional 20−30 cycles, a very large amount of the products shown in (**4**) is produced. (**B**) The 'nested primer' variation of PCR (reference 22). After the first amplification by the 'outside' set of oligonucleotides shown in (**A**), a small amount of the products in (**A**) are used as template in a second amplification reaction, this time using amplified DNA fragment. The same type of reaction, with 20−30 cycles, is performed, producing several microgrammes of the specific DNA fragment from an initial few hundred nanogrammes of genomic DNA in the first amplification. This second amplification often increases the specificity of the PCR procedure.

It is probably for this reason that very long stretches (>2000 bp) of sequence can be amplified routinely by PCR with the heat-resistant polymerase.

4.1 **Materials**

The following buffers and reagents are needed for performing the PCR reaction as described in this section.

(i) *10 × PCR buffer.* For 1 ml use the following:

670 mM Tris−HCl, pH 8.8	670 μl of 1 M
67 mM MgCl$_2$	67 μl of 1 M
166 mM ammonium sulphate	83 μl of 2 M
100 mM β-mercaptoethanol	7 μl of 14 M
	177 μl of water

(ii) *dNTP stock solutions.* These are four separate solutions of each deoxyribonucleotide triphosphate at 50 mM in water. Store at −70°C in small aliquots. We have found that the concentrated solutions of dNTPs sold by Pharmacia are of reproducible high quality.

(iii) *10 × dNTP mix.* For 0.2 ml use the following:

12.5 mM dATP	50 μl of 50 mM dATP stock
12.5 mM dTTP	50 μl of 50 mM dTTP stock
12.5 mM dGTP	50 μl of 50 mM dGTP stock
12.5 mM dCTP	50 μl of 50 mM dCTP stock

(iv) *Dimethyl sulphoxide.* Any high grade anhydrous dimethyl sulphoxide (DMSO) will suffice.

(v) *TE buffer.* This buffer is used to prepare all DNA samples described in this chapter. For 100 ml:

10 mM Tris−HCl, pH 7.8	1 ml of 1 M
1 mM EDTA, pH 8.0	0.2 ml of 0.5 M
	99 ml of water

(vi) *Template DNA.* In our experience, genomic DNA extracted by a variety of methods can be used to amplify regions by PCR (see, for example ref. 24). After extraction of genomic DNA, it should be prepared as a 100 μg/ml stock solution; it need not be digested with restriction enzymes. To prepare cloned DNA samples for PCR, digest plasmid DNA with a restriction enzyme that cuts outside the region to be amplified to relieve the torsional strain of supercoiling. After removing the restriction enzyme by phenol extraction and ethanol precipitation, resuspend the cloned DNA samples in TE buffer at 100 μg/ml. A very dilute solution of the cloned DNA sample (~0.01 μg/ml) is prepared by diluting the sample in TE with carrier tRNA (see Section 5.1) at 100 μg/ml.

(vii) *Oligonucleotide primers.* Oligonucleotide primers should be 20−25 nucleotides in length to ensure their relative uniqueness in the genome. The primers should be gel-purified following synthesis and stored in TE buffer at −20°C at 10 pmol/μl.

(viii) *Taq DNA polymerase.* Heat-stable DNA polymerase purified from the thermophilic bacterium *T.aquaticus* can be purchased from several commercial sources, including New England Biolabs, (Beverly, MA) and Perkin-Elmer, Inc. The enzyme from these sources is usually between 1 and 10 U/μl.

4.2 Methods

(i) Combine the following reagents in a microcentrifuge tube:
 5 μl of 10 × PCR buffer
 5 μl of 10 × dNTP mix
 5 μl of DMSO
 1.5−5 μl of genomic DNA (100 μg/ml)
 or
 1.5−5 μl of restriction-digested cloned DNA (0.01 μg/ml)
 25 pmol of each oligonucleotide primer (only the first 'outside' pair if the nested oligo procedure is used).
 Add water to 50 μl.

(ii) Incubate the mixture at 95°C for 5 min to denature the double-stranded DNA.

(iii) Spin the tube briefly in a microcentrifuge to collect all the condensate at the bottom of the tube.

(iv) Add 2 units of *Taq* DNA polymerase to the sample.

(v) Layer 50 μl of mineral oil on top of the aqueous sample to prevent condensation during the subsequent heating reactions.

(vi) Incubate the sample at 50°C for 30 sec to allow the oligonucleotides to anneal to their complementary sequences in the DNA.

(vii) Incubate the sample at 65°C to allow the DNA polymerization to proceed. The length of time required for complete polymerization depends on the length of the DNA sequence being amplified. A rough estimate of the rate of polymerization under these conditions is 500 nucleotides per minute; the minimum time for incubation is 1 min.

(viii) Incubate the sample at 93°C for 1 min to denature the double strands.

(ix) Repeat steps (vi)−(viii) 25−30 times. It is not necessary to add *Taq* DNA polymerase between cycles.

(x) If the nested oligo procedure is used, transfer 1 μl of a 1:10 dilution of the first amplification reaction to another tube that contains the following:
 5 μl of 10 × PCR buffer
 5 μl of 10 × dNTP mix
 5 μl of DMSO
 25 pmol of each of the second internal pair of oligonucleotides
 Water to 50 μl.

(xi) Repeat steps (ii)−(ix) to amplify the second internal sequence.

(xii) Analyse about one-tenth of the amplified product by agarose gel electrophoresis. In many cases, the amplification reaction produces a single ethidium-staining band. It is possible to roughly quantitate the amount of DNA in the amplified fragment from the agarose gel. Generally, amplification of a 1-kb DNA sequence from 0.3 μg of mammalian genomic DNA results in about 1 μg of the amplified sequence. When a very small fragment of the amplified DNA from the first set of oligonucleotides is amplified with a second internal set of oligonucleotides as described above, several microgrammes of the specific internal DNA fragment are produced.

In some cases, additional ethidium-staining bands, whose sizes are different from that expected, are seen on the gels. The origin of these DNA species is not known. However, they do not appear to interfere with the analysis of the specific fragment by RNase cleavage or DGGE. In most cases, these additional bands do not appear in the second set of amplifications when the nested oligo method is used.

5. THE RNase CLEAVAGE PROCEDURE

The experimental details of each of the steps of the RNase cleavage reaction, including the method used to synthesize the labelled RNA probe, are outlined below. For the most part, the steps are very similar to the original publication; however, some improvements have been incorporated into the present discussion. The protocol also discusses some trouble spots and possible ways to overcome them.

Note that the RNA probe in this procedure need not correspond exactly to the complementary sequence in the test DNA sample at its ends. Non-complementary overhangs at the 5' and/or 3' end of the labelled probe do not interfere with the analysis; they are simply destroyed by the RNase during the cleavage reaction. In fact, it is useful to have overhangs that make the probe 5−10% longer than the test DNA fragment. This arrangement allows the full-length probe to be distinguished from the full-length wild-type hybrid fragment resulting from RNase cleavage (see *Figure 1*). This test is one way of assuring that the RNase is active. With too little RNase, or degraded RNase, the overhangs are not removed efficiently, and this can be monitored on the autoradiogram.

5.1 **Materials**

5.1.2 *Reagents for generating the RNA probes*

The following reagents are needed for generating uniformly-labelled RNA probes that are used in the RNase cleavage procedure. Conditions are given for using RNA polymerases and promoters from SP5, T7 and T3 bacteriophages. For further information about synthesizing RNA probes, the reader is referred to the original SP6 publications (35,26) and to a recent methods paper (37).

In order to avoid problems with stray ribonucleases, it is best to exercise some degree of care in preparing reagents and handling tubes, pipettors, etc. Solutions that are heat stable should be autoclaved and kept sterile; solutions, such as bovine serum albumin (BSA), dithiothreitol (DTT), etc. that cannot be autoclaved should be made with sterile water and stored in sterile tubes or bottles. It is often helpful to use autoclaved plasticware and to clean the stems of pipettors frequently when working with RNA.

(i) *Template DNA.* The target DNA fragment must be cloned into a plasmid or phage vector adjacent to one of the highly specific bacteriophage promoter sequences. The three systems that are routinely used are phages SP6, T7 and T3. Several vectors are available with one or two of the promoters facing toward a polylinker cloning site [examples: pSP70 series of Krieg and Melton (39); pGEM series of Promega Biotec, Inc. (Madison, WI); pBluescribe series of Stratagene, Inc. (La Jolla, CA)]. Ideally, the target sequence is cloned into the polylinker between two promoters so that run-off RNA transcripts can be made from each strand of the target sequence.

   After cloning the target sequence into a vector containing two different promoters, digest $10-20$ μg of the plasmid DNA with a restriction enzyme that cleaves at one end of the target sequence (or within the polylinker sequence) to prepare the template for run-off transcription. If both strands of the template are to be used as probe, also digest $10-20$ μg of the DNA with a different restriction enzyme that cleaves at the opposite end of the target sequence. Extract the digested DNA with phenol, precipitate with ethanol and resuspend the DNA in TE to 1000 μg/ml.

(ii) *RNA polymerases.* Purified SP6, T7 and T3 RNA polymerases are available from several commercial sources. Generally, they are sold at concentrations of $5-20$ U/ml.

(iii) *Labelled nucleotide.* Purchase one of the four nucleotide triphosphates (NTPs) labelled in the alpha position with ^{32}P at a specific radioactivity of 400 Ci/mmol. We recommend the use of GTP as the labelled nucleotide, and the procedure below describes its use for making probes. However, one of the other NTPs can also be used as the labelled compound, with corresponding alteration of the amounts of the remaining unlabelled NTPs used.

(iv) *Cold GTP.* A stock solution of cold GTP (or the nucleotide corresponding to the labelled NTP) at 2 mM. Store at $-70°$C.

(v) *3-NTP mix.* Stock solutions of the other three NTPs, at 50 or 100 mM, are used to prepare a '3-NTP mix', which is a solution of all three nucleotides, each at 10 mM in TE. Store at $-70°$C.

(vi) *$10 \times$ transcription buffer.* For 1 ml use the following;

400 mM Tris$-$HCl, pH 7.5	400 μl of 1 M
60 mM MgCl$_2$	60 μl of 1 M
20 mM spermidine	200 μl of 100 mM
	340 μl of water

(vii) *DTT stock.* Make a stock solution of 1.0 M DTT in water and store at $-20°$C.

(viii) *BSA stock.* Prepare a solution of RNase-free BSA at 1 mg/ml in water and store at $-20°$C.

(ix) *Placental RNase inhibitor.* Placental RNase inhibitor is available commercially from several sources, including Promega Biotec,Inc.

(x) *DNase I.* Purchase ultra-pure, RNase-free DNase I and prepare a solution at 1 mg/ml in water; store in small aliquots at $-20°C$ or $-70°C$. We have had good results with DNase from Cooper Biomedical (Worthington; catalogue LS06333).

(xi) *SDS.* Prepare a stock of 25% sodium dodecyl sulphate (SDS) in water. Use an 'ultrapure' grade if possible. Do not autoclave. Store at room temperature.

(xii) *Carrier tRNA.* For several steps in the procedure, RNase-free and DNase-free carrier RNA is needed. We recommend the use of yeast tRNA, which should be meticulously cleaned by the following procedure.

(a)   Suspend the tRNA at 20 mg/ml in TE, add SDS to 0.1%, and then add proteinase K to 100 μg/ml.
(b)   Incubate at $37-50°C$ overnight.
(c)   Extract with phenol three or four times, and dialyse the tRNA against several litres of TE overnight.
(d)   Adjust the concentration to 5 mg/ml and store in aliquots at $-70°C$.

5.1.2 *Reagents for annealing, cleavage and gel analysis*

The following reagents are needed for the annealing and cleavage reactions, and for gel analysis for the RNase cleavage procedure.

(i) *Test DNA.* The DNA to be tested can be genomic or cloned DNA amplified by PCR, cloned DNA or unamplified genomic DNA. DNA prepared by PCR is described in Section 4. Cloned DNA and unamplified genomic DNA should be digested with a restriction enzyme prior to use so that accurate quantitation and complete denaturation and annealing to the probe can be performed. Digest the DNA with restriction enzymes that cleave outside the region to be probed, remove the enzymes by phenol extraction and concentrate the DNA by ethanol precipitation. Re-suspend the DNA in TE buffer to a concentration of 100 μg/ml.

(ii) *Hybridization buffer.* For 10 ml use the following:

80% formamide	8 ml of de-ionized formamide
50 mM Pipes buffer, pH 6.4	1 ml of 0.5 M Pipes, pH 6.4
500 mM NaCl	1 ml of 5 M NaCl
1 mM EDTA	20 μl of 0.5 M EDTA, pH 8.0
	Water to 10 ml
	Do not autoclave

Note: purchase a high-quality grade of formamide and de-ionize by stirring gently with Dowex AG50W-X8 (200−400 mesh; 'Mixed Bed Resin' from Biorad, Inc.) or its equivalent for 30 min at room temperature; use about 2 g resin/100 ml formamide.

116

Collect the formamide by filtering it through Whatman 1MM paper, and store in tightly capped tubes at $-20°C$ or $-70°C$.

Some batches of formamide contain contaminants that appear to degrade RNA, especially at high temperatures; this causes high background on the autoradiograms following the RNase cleavage reaction. If problems with background persist, we recommend that formamide be re-crystallized prior to use.

(iii) *RNase reaction buffer.* For 100 ml use the following:

20 mM Tris−HCl, pH 7.5	2 ml of 1 M Tris−HCl, pH 7.5
200 mM NaCl	4 ml of 5 M NaCl
100 mM LiCl	10 ml of 1 M LiCl
1 mM EDTA	0.2 ml of 0.5 M EDTA, pH 8.0
	Water to 100 ml
	Autoclave to sterilize.

(iv) *RNase A stock solution.* Purchase purified ribonuclease A (RNase A) in the crystallized form. Our best results have been obtained with a preparation sold by Sigma (Catalogue No. R-5125), although RNase A from other commercial sources has been used successfully also.

(a)   In a screw-cap tube, prepare a solution of 2 mg/ml in water, being very careful to avoid contamination of hands, laboratory bench, etc., since RNase A is extremely stable and active.

(b)   Cap the tube tightly, and place it in a beaker of boiling water for $10-15$ min to destroy any traces of DNase.

(c)   Cool to room temperature and store at $-20°C$.

(v) *Proteinase K.* Crystalline proteinase K can be purchased from several commercial sources, including Boehringer Mannheim. Make a stock solution at 10 mg/ml in 100 mM Tris−HCl, pH 7.5, 10 mM EDTA and 1% SDS. Store in aliquots at $-20°C$.

(vi) *2FC.* 2FC is a mixture of chloroform, phenol and isoamyl alcohol that is used to extract protein from the RNase reaction. For 500 ml:

50% phenol (equilibrated to pH 7 with TE)	250 ml
48% chloroform	240 ml
2% isoamyl alcohol	10 ml
0.01% hydroxyquinoline	50 mg

Store at 4°C in a dark bottle.

(vii) *Formamide loading solution.* For 10 ml:

80% de-ionized formamide	8 ml
1 × TBE	1 ml 10 ×

0.01% xylene cyanol	0.1 ml 1%
0.01% bromophenol blue	0.1 ml 1%
	Water to 10 ml

Store at −20°C.

10 × TBE = 0.9 M Tris base, 0.9 M boric acid, 25 mM EDTA; pH adjusted to 8.3.

(viii) *Sequencing gels.* The labelled RNA products resulting from the RNase cleavage reactions are analysed by separating the protected RNA strands from their complementary DNA strands, followed by electrophoresis on denaturing (urea) polyacrylamide gels of the type used for DNA sequencing. Preparation and running of these gels is simple and well-described elsewhere (24).

The percentage acrylamide used to analyse the RNA products from the cleavage reaction should be chosen to maximize the resolution of the products. As a general rule, if the RNA probe is 500 nucleotides or shorter, a 10 or 12% polyacrylamide gel should be used. For longer RNA probes, 500−1000 nucleotides, a 5% or 6% polyacrylamide gel should be used.

5.2 Methods

5.2.1 RNA probe synthesis

The method below describes the synthesis of RNA probes by *in vitro* transcription of cloned DNA templates carrying a bacteriophage promoter; this method is essentially that originally described in references 37 and 38. Note that the specific radioactivity of the RNA probes that are made by the following method is lower than that which we originally described for testing genomic DNA samples. These lower specific radioactivity probes, which contain about 10 Ci/mmol in one of the labelled NTPs, are sufficiently radioactive to use with one tenth of the total amount of DNA resulting from the PCR reaction described in Section 4, or with a very small amount of cloned DNA. The advantages of using these low specific radioactivity probes are that they are stable for over 1 week (versus 1−2 days for the high specific radioactivity probes) and that it is easier to synthesize full-length probes because NTPs do not become limiting in the transcription reaction. If PCR is not used, and genomic DNA is to be tested, high specific radioactivity probes should be synthesized, using 200−400 Ci/mmol in one of the labelled NTPs. It is important that the concentration of the labelled NTP is at least 10 μM at the beginning of the transcription reaction. Therefore, high specific radioactivity probe synthesis requires the use of 40−80 μCi of labelled NTP.

(i) Combine the following components in a sterile microcentrifuge tube, in the order listed:

9 μl of water
1 μl of restriction-digested template DNA (1 mg/ml)
1 μl of 3-NTP mix
1 μl of cold GTP (2 mM)
1 μl of [α-^{32}P]GTP (10 mCi/ml; 400 Ci/mmol)
2 μl of DTT (1.0 M)
2 μl of BSA (1 mg/ml)

2 – 5 units of placental RNase inhibitor

2 μl of 10 × transcription buffer

(ii)   Add 3 – 5 units of SP6, T7 or T3 RNA polymerase to the mixed sample.

(iii)   Incubate at 40°C for 1 h to allow transcription to proceed.

(iv)   Add 1 μl of DNase I (1 mg/ml) to destroy the template DNA; incubate at 37°C for 20 min.

(v)   Add 10 μg of carrier tRNA, 60 μl of water and 20 μl of ammonium acetate (7 M) to the sample, and extract it once with 2FC.

(vi)   Precipitate the extracted aqueous layer with ethanol; re-suspend the pellet in 80 μl of water and 20 μl of ammonium acetate (7 M) and precipitate with ethanol agin.

(vii) Re-suspend the RNA pellet in 200 μl of TE buffer containing 0.1% SDS. Store at −20°C.

5.2.2 Annealing and cleavage reactions

In our original method, a molar excess of labelled RNA probe was used to drive the hybridization reaction between the probe and the small quantity of complementary DNA sequences in the several microgrammes of total genomic DNA in the test sample (16). Now, however, because such a large molar amount of specific DNA fragment can be made from amplification of genomic DNA by PCR, it is no longer necessary to use excess probe; in fact, the experiments below use excess test DNA. This modification improves the signal-to-noise ratio since it is not necessary to remove residual full-length probe by RNase digestion.

When setting up an experiment, it is useful to include the following controls.

Control 1: a mock hybridization of the probe to carrier tRNA.

Control 2: a positive wild-type control of the probe annealed to the cloned wild-type DNA fragment.

Control 3: the probe annealed to a known cloned or PCR-genomic mutant DNA fragment that results in an efficiently-cleaved mismatch, if available.

(i)   Combine the following in a sterile microcentrifuge tube:

        1 μl of PCR-amplified genomic DNA (~ 100 μg/ml; from either one set of oligonucleotides or the second set if the nested oligo method is used) *or* 1 μl of restriction-digested cloned DNA (100 μg/ml).

        1 μl of labelled RNA probe

        30 μl of hybridization buffer.

(ii)   Heat the mixture at 95 – 100°C for 10 min in a boiling water bath.

(iii)   Briefly spin the sample in a microcentrifuge to collect the condensate at the bottom of the tube.

(iv)   Incubate the mixture at 45 – 50°C for 60 min to allow the probe to anneal to its complementary sequences in the test DNA.

        Note: our original procedure designated a hybridization time of 10 – 12 h to allow complete annealing of the genomic DNA fragments to the probe. With the quantities of cloned and PCR-amplified genomic DNA fragments, the concentration of target DNA is much higher than with genomic DNA, so the hybridization times can be decreased dramatically. These shorter incubation times make the procedure more convenient to perform and also decrease the degree of degradation of the probe.

(v)   Add 350 μl of RNase reaction buffer/RNase mix to the annealed sample. Mix well by vortexing.

The RNase reaction buffer/RNase mix is: RNase reaction buffer with RNase A (boiled) at a final concentration of 40 μg/ml and with tRNA carrier at 20 μg/ml. This solution can be prepared a few hours before use and stored on ice. For 5 ml (sufficient for 13−14 reactions): 5 ml of RNase reaction buffer, 100 μl of boiled RNase A stock solution (2 mg/ml), 20 μl of tRNA (5 mg/ml).

Note: most experiments that we have performed with RNase cleavage have used RNase A at 40 μg/ml. However, we found that with one batch of the enzyme, 40 μg/ml was far too much; the RNA:DNA hybrids were damaged with this quantity. We were able to get good results with that batch of enzyme by decreasing the RNase A concentration to 4 μg/ml. For this reason, we recommend that each new batch of RNase A be titrated over the concentration range of 2−40 μg/ml. If available, both a wild-type and mutant control test DNA sample should be used in the titration to determine the optimal RNase A concentration that results in efficient cleavage at the mismatch in the mutant sample and intact, low background signals in the wild-type sample.

(vi)   Incubate at 25°C for 60 min.

Note: our original method used a 30 min incubation for the RNase reaction. Under those conditions, many mismatches were cleaved by RNase A only partially, resulting in a radioactive signal appearing at the wild-type position in the gel. We subsequently found that longer digestion times would drive most of the partially-cleaved mismatches to complete cleavage (15). In fact, it is often easiest to interpret cleavage results when a time course of RNase treatment is performed, by removing aliquots at 30, 60 and 90 min of incubation. For some mismatches, 90 min of RNase treatment is a little too harsh, resulting in additional degradation at the ends of the RNA fragments. Therefore, if a time course is not performed, the best single time of treatment is about 60 min.

(vii)   Stop the RNase reaction by adding 10 μl of SDS (25%) and 10 μl of proteinase K (10 mg/ml). Incubate at 37°C for 30 min.

(viii)   Add 10 μg of carrier tRNA.

(ix)   Extract the reaction once with phenol and once with 2FC. In the first extraction, carefully remove only 300 μl of the aqueous supernatant with a plastic pipette tip and transfer it to another tube. This precaution prevents the transfer of residual RNase from the interface between the phenol and the aqueous layer.

(x)   Precipitate the nucleic acids with ethanol. Rinse the pellet with 70% ethanol and dry.

(xi)   Re-suspend the RNA pellet in 10−20 μl of formamide loading solution. It is often difficult to completely re-dissolve the RNA in the formamide solution; the pellet can be more quickly dissolved by pipetting in and out several times with a plastic pipette tip and a pipettor.

5.2.3 *Analysis*

(i)   Pre-electrophorese the sequencing gel for 1−2 h prior to loading.

(ii)   Place the re-suspended samples at 95 – 100°C for 4 min to separate the RNA strands from the DNA strands.

(iii)   Turn off the power supply and disconnect the electric leads.

(iv)   Immediately before loading a lane on the gel, swish it out with a drawn-out capillary pipettor or a hypodermic needle and syringe.

(v)   Remove a sample from the hot bath and quickly load 5 µl of it in a lane on the gel. Continue quickly with each sample until all are loaded.

(vi)   Connect the electric leads to the gel apparatus and turn the power supply back on. Electrophorese at 15 – 20 W until the bromophenol blue dye runs off the bottom of the gel.

(vii) At the end of the run, turn off the power supply, disconnect the leads and remove the gel plates from the apparatus. Dry the gel onto a sheet of Whatman 1MM paper or its equivalent and expose the dried gel to X-ray film.
Note: the gel can be exposed to X-ray film without drying if desired. Drying the gel increases the resolution and sensitivity somewhat, and makes it easier to obtain multiple exposures of the gel.

6. DENATURING GRADIENT GEL ELECTROPHORESIS

This section describes the DGGE system in detail.

6.1 The gel system

Most naturally-occurring DNA sequences melt in the temperature range between 60°C and 90°C. In order to achieve these temperatures and to establish a relatively steep denaturing gradient in the gels, it is necessary to run the gels at a high temperature near the melting temperatures of naturally-occurring DNA. A temperature of 60°C is generally used. This is best achieved by running the gels submerged in a bath heated to 60°C. A special apparatus designed by L.Lerman and S.Fischer is the system that we use routinely, and is described in detail in references 20 and 28. The entire system consists of:

(i)   a gel apparatus that holds gel plates in place so that an upper electrophoresis chamber is created;

(ii)   an aquarium that serves as the heated bath and lower electrophoresis chamber;

(iii)   a heater/stirrer;

(iv)   electrodes;

(v)   a peristaltic pump to continually replace the buffer in the small upper chamber from the large lower chamber.

This system maintains the temperature of the gels during electrophoresis to within 0.1°C.

6.1.1 Gel equipment

Several options exist for obtaining a suitable gel apparatus and heating system for DGGE, including purchasing the gel apparatus from several companies or building it from the drawings in references 20 and 28. The following vendors offer either part or all of the equipment needed to run denaturing gradient gels.

(i)   Green Mountain Lab Supply, 86 Central St, Waltham, MA 02154, USA. This company sells a gel box made to the specifications of the design in references 20 and 28, as well as selling the Plexiglas aquarium, heating/stirring apparatus, glass plates, combs, spacers, peristaltic pump and the electrodes.

(ii)   Hoeffer Scientific Instruments. This company sells a gel apparatus that can be adapted for DGGE. The apparatus contains a gel frame that holds glass plates vertically, and has an upper electrophoresis chamber that attaches to the top of the gel plates. The manufacturers provide instructions for setting up the gel plate and pouring the gel. After pouring the gel, the chamber is attached, and the set-up is mounted to the aquarium so that the gel is submerged in the running buffer. As usual, the buffer is circulated through the upper chamber by peristaltic pumping. The catalogue number for this apparatus is Hoeffer SE600.

(iii)   CBS Scientific Co., PO Box 856, Del Mar, CA 92014, USA. The apparatus sold by this company is similar to that listed in (ii).

It is also possible to build the Plexiglas gel apparatus relatively inexpensively. The photographs in references 20 and 28 may be sufficient. If desired, detailed plans can be obtained from R.M.Myers.

In addition to the gel apparatus, the system requires several other components that are available in many laboratories. These components and the vendors from which our laboratories have purchased some of them are listed below.

(i) *Heating/stirring apparatus.* Example: Model 70, PolyScience Corporation (Niles, IL).

(ii) *Peristaltic pump.* Catalogue No. J7543-20 (pump motor), No. J7017-20 (pump head), and No. J6411-17 (silicon tubing) from Cole Parmer, Inc. (Chicago, IL).

(iii) *Aquarium.* An aquarium of dimensions 22 inch wide × 6 inch deep × 10 inch tall, made of Plexiglas or 1/4 inch thick glass is most convenient. This aquarium will hold up to two gel boxes. An aquarium can easily be constructed from Plexiglas pieces (glued together with water- and heat-resistant epoxy) or glass pieces (glued together with silicon cement, which is often used for fish aquaria). Green Mountain Lab Supply also sells an aquarium built specifically for DGGE. Alternatively, aquaria can be purchasd from pet supply stores; the only precaution is that the aquarium should have no metal parts.

(iv) *Power supply.* Any power supply capable of providing 150 V and 150 mA is sufficient.

(v) *Electrodes.* A platinum anode, which is immersed into the aquarium buffer, can be made as described in reference 20. The aquarium sold by Green Mountain Lab Supply has a built-in anode. The cathode, which is placed across the small upper chamber of the gel apparatus during electrophoresis, is constructed of either graphite or platinum. Both Green Mountain Lab Supply and CBS Scientific Co. sell graphite cathodes.

6.1.2 *Materials*

(i) *Acrylamide stock solution.* For 250 ml use the following:

40% acrylamide	100 g of acrylamide
1.07% bisacrylamide	2.7 g of bisacrylamide
	Water to 250 ml
	Do not autoclave.

Both reagents should be of the highest purity available. Note that this solution contains a 37.5:1 ratio of acrylamide to bisacrylamide, which results in about half the degree of cross-linking that is used in typical sequencing gels.

(ii) *20 × gel running buffer.* For 2 litres use the following:

800 mM Tris base	190 g of Tris base
400 mM sodium acetate	110 g of NaAc × 3 water
20 mM EDTA	15 g of Na_2EDTA
pH 7.4	Water to about 2 litres
	Adjust the pH to 7.4 with acetic acid.

(iii) *80% denaturant stock solution/7% acrylamide.* For 250 ml use:

7% acrylamide (37.5:1)	44 ml of acrylamide stock solution
32% formamide	80 ml of de-ionized formamide (see Section 5.1)
5.6 M urea	85 g of urea (ultrapure)
1 × gel running buffer	12.5 ml of 20 × gel running buffer
	Water to 250 ml

(iv) *0% denaturant stock solution/7% acrylamide.* For 250 ml use:

7% acrylamide (37.5:1)	44 ml of acrylamide stock solution
1 × gel running buffer	12.5 ml of 20 ×
	Water to 250 ml

Note: for most test DNA fragments, acrylamide concentrations of 7% are best for DGGE. However, in some cases, particularly when DNA fragments smaller than 300 bp are being examined, better resolution can be obtained with a higher acrylamide concentration, such as 10% or 14%. In these cases, the 80% and 0% denaturant stock solutions should be made up with accordingly higher acrylamide concentrations.

Store all denaturant stock solutions at +4°C in dark bottles.

(v) *Polymerization catalysts.* TEMED (*N,N,N',N'*-tetramethylethylenediamine). For 50 ml of ammonium persulphate stock (10%) use 5 g of ammonium persulphate and water to 50 ml. Store at +4°C.

(vi) *Neutral loading solution.* For preparation of 10 ml:

20% Ficoll or sucrose	2 g of Ficoll 400 or sucrose
10 mM Tris−HCl, pH 7.8	100 μl of 1.0 M Tris−HCl, pH 7.8
1 mM EDTA	20 μl of 0.5 M EDTA, pH 8.0
0.1% dye (orange G, xylene cyanol, or bromophenol blue	10 mg of dye
	Water to 10 ml

6.1.3 *Preparation of perpendicular gradient gels*

As discussed in detail below (Section 6.2), gels containing a gradient of denaturants perpendicular to the direction of electrophoresis are used to determine the melting behaviour of a new DNA fragment. Such gels provide the information needed to plan the optimal gradients and electrophoresis times for normal parallel gradient gels. This section describes the procedures for pouring such gels. Additional details of these gels are discussed in reference 20.

A perpendicular gradient gel consists of an acrylamide gel containing a linearly-increasing gradient of DNA denaturants, from left to right, and a single large 'well' at the top of the gel into which the sample is loaded. These gels are prepared by setting one of the side spacers (the left side) between the plates so that a 3−4 cm gap is made at the bottom of the gel (see *Figure 7*). In addition, a single large spacer is set at the top of the gel, and the joints between it and the two top spacers are sealed with a small amount of vacuum grease. After clamping the plates into the gel apparatus and sealing the bottom and sides with 1% agarose (in 1 × gel running buffer), the apparatus is turned on its side and the gradient mixture is poured through the 3−4 cm opening on the left side (see *Figure 7*). The total volume of acrylamide solution required to fill the gel should be determined empirically; most gel boxes and spacers require a total volume of 16−24 ml.

To determine the melting properties of a new DNA fragment, the electrophoretic behaviour of the fragment is measured in a perpendicular denaturing gradient gel, containing a fixed acrylamide concentration with 0% denaturant on the left side and 80% denaturant on the right side. For a set of plates that holds a total of 16 ml of acrylamide, such a gel is prepared as follows.

(i)   Add 5 μl of TEMED and 80 μl of 10% ammonium persulphate to 8 ml of 80% denaturant stock solution, mix well and place on ice.

(ii)   Add 5 μl of TEMED and 80 μl of 10% ammonium persulphate to 8 ml of 0% denaturant stock solution, mix well and place on ice.

(iii) Pour the 80% solution into the side of the gradient maker that exits into the tubing that feeds into the gel. Remove any air bubbles that are present between the channel that connects the two chambers of the gradient maker.

(iv)   Pour the 0% denaturant stock solution into the other chamber of the gradient maker.

(v)   While mixing the 80% side, pour the gradient solution into the gel apparatus by letting it dribble slowly into the gel plates through the 1−2 in gap on the left side of the apparatus. Thus, a concentration of 80% denaturant will land at the bottom of the apparatus sitting on its side, and a concentration of 0%

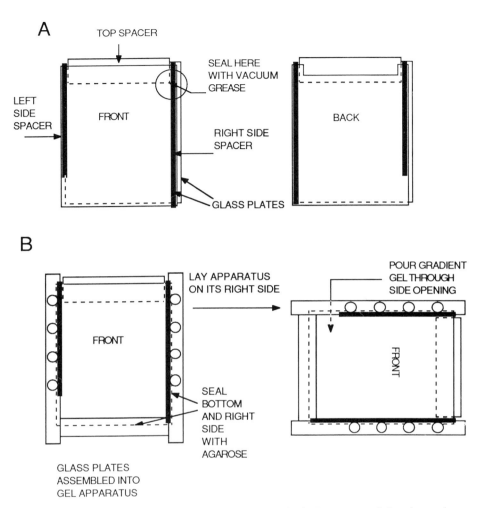

Figure 7. Preparation of a perpendicular denaturing gradient gel. (**A**) Arrangement of glass plates and spacers. The left drawing is a front view of the glass plates assembled together with the spacers arranged for perpendicular gel analysis. From this view, the front plate is rectangular (not earred) and the back plate is earred. The spacer on the left side is placed so that a 3–4 cm gap appears at the bottom of the plates. The right spacer is placed so that it extends from the top to the bottom of the gel. The top spacer is placed so that it extends about 1 cm below the top of the earred plate and forms a tight joint between the two side spacers. The joint on the right side of the top spacer is sealed with a small amount of vacuum grease. The right drawing shows a back view of the same assembled glass plates. (**B**) Assembled glass plates and gel apparatus. The assembled glass plates and spacers in (**A**) are clamped into the gel apparatus, and the bottom and right side of the gel plates are sealed with molten agarose. The bottom seal is made by pouring about 50 ml of agarose into a trough in which the plates sit. After the agarose hardens, the gel apparatus is placed on its right side and the gradient gel is poured through a hole in the gel apparatus and the 3–4 cm space at the left hand side of the gel.

denaturant will exist at the top of the apparatus sitting on its side after the gel is poured. When the acrylamide polymerizes, the gel is placed in its normal upright position, and the denaturing gradient goes from 0% to 80% from left to right.

125

(vi)   After polymerization, stand the gel in its upright position and remove the top spacer.
(vii)   Place the gel apparatus into the heated gel running buffer in the aquarium and connect the peristaltic pump so that the buffer runs from the lower chamber into the upper chamber and flows over the top of the apparatus.
(viii) Follow the instructions in Section 6.2.1 for preparing and loading the DNA samples on the perpendicular gels.

6.1.4 *Preparation of parallel gradient gels*

Parallel gels contain a linearly-increasing gradient of DNA denaturants, parallel to the direction of electrophoresis, from the top of the gel to the bottom of the gel. The denaturant range over which the gradient extends is chosen from the information gained from perpendicular gels about the melting behaviour of the test fragment. Generally, the range is about 25%, established so that the test DNA fragment begins to melt about in the middle of the gel. For example, a parallel gel of 25−50% denaturant would be used to analyse a DNA fragment with a domain that melts at 37.5% denaturant. Parallel gels contain lanes across the top so that multiple samples can be tested in a single gel.

Parallel gels are prepared and run in a manner similar to perpendicular gels. The two spacers are placed on the sides from top to bottom in a parallel gel, in the same way spacers are set for sequencing or protein gels. After clamping the plates into the gel apparatus, the bottom and sides of the plates are sealed with agarose as usual. The gel is poured in the same way that perpendicular gels are poured, except that the gel solution is fed into the plates from the top of the apparatus. After pouring, a Delrin or Teflon comb is inserted into the gel and the gel is left to polymerize. After polymerization, the comb is removed and the apparatus is set into the heated aquarium and connections are made as described for perpendicular gels. Sample preparation and electrophoresis of parallel gels is described in Section 6.2.2.

6.2 **Preliminary experiments: perpendicular and travel schedule gels**

To determine the optimal conditions for analysing a DNA fragment for base changes by DGGE, two relatively simple preliminary experiments are performed.

(i)   The analysis of the fragment on a perpendicular gel to determine its melting domain structure and the Tm(s) of its domains. The information from this experiment defines the optimal gradient conditions for multilane analysis in parallel gels.
(ii)   A 'travel schedule gel' which is a parallel gel, containing an optimal range of denaturant as determined by perpendicular gel analysis, and containing multiple loadings of the test sample run for various lengths of time. This type of gel, though not always necessary, provides an indication of the optimal length of time to run a parallel gel for maximum resolution of base changes in test samples for a particular DNA fragment.

6.2.1 *Perpendicular gels: determination of melting behaviour*

In order to obtain some indication of the way that a DNA fragment melts in solution, several approaches can be taken. If the DNA sequence of the fragment is known, it is possible to predict the number, locations and Tms of melting domains in the fragment, by use of a computer algorithm developed by L.Lerman and his colleagues (38,39; see also 18,19). This information can be used to calculate the gradient conditions and electrophoresis time that will result in maximum resolution of mutant fragments in the gel.

Although the computer algorithm that predicts melting behaviour is accurate and powerful, it is possible instead to use an empirical determination to obtain enough information about melting behaviour to optimize gel conditions. This test involves electrophoresis of the DNA fragment in a perpendicular denaturing gradient gel. This method is particularly important when testing anonymous DNA fragments for polymorphisms and in other cases where the nucleotide sequence of the DNA fragment is not known.

Perpendicular gel analysis of a DNA fragment gives an indication of the relationship between the mobility of the fragment in polyacrylamide as a function of denaturant concentration. The DNA sample is layered across the top of the gel, which contains a gradient of denaturants increasing from left to right. As electrophoresis occurs, DNA fragments on the left, low denaturant concentration side of the gel travel through the gel matrix with the rapid, linear mobility of double-stranded DNA. The gradient conditions are chosen so that these fragments never undergo even partial melting throughout the run. On the right side of the gel, however, where the denaturant concentration is highest, the DNA fragments begin to melt as soon as they enter the gel, causing an abrupt decrease in mobility. At other positions of the gel, where the denaturant concentration is intermediate, varying degrees of melting occur, and intermediate mobility retardations are seen. A steep transition in mobility, which is seen as an increase in the slope of the line formed by the DNA fragment, occurs at a denaturant concentration corresponding to the Tm at which the cooperative melting of a domain occurs. The number of mobility transitions is equal to one less than the number of melting domains in a fragment, since the fragment comes apart and does not show a steep retardation when the last domain melts.

In *Figure 8A*, a DNA fragment with a single transition, and therefore two melting domains, gives a steep transition whose mid-point is at a position in the gel where the denaturant concentration is equal to about 58%; this position corresponds to the Tm of the lowest temperature melting domain of the fragment. From this information, one can choose the optimal gel conditions to examine this DNA fragment on parallel gels; the best resolution occurs with about 10−15% denaturant on either side of the Tm, so a parallel gel of 45−70% would be chosen. In this type of parallel gel, the DNA fragment enters the gel as a completely double-stranded molecule until it migrates into 38% denaturant, where the first domain melts and the fragment undergoes a mobility reduction.

A second fragment, with two transitions on the perpendicular gel (*Figure 8B*), contains three melting domains. The lowest two domains are seen in the two transitions on the gel; the first melts at 38% denaturant and the second melts at 68% denaturant. Because these two domains are far apart, it is best to analyse each one separately on two different parallel gels. The first is examined on a 25−50% parallel gel, and the second domain is examined on a 55−80% gel. In the first parallel gel, similar to the fragment analysed

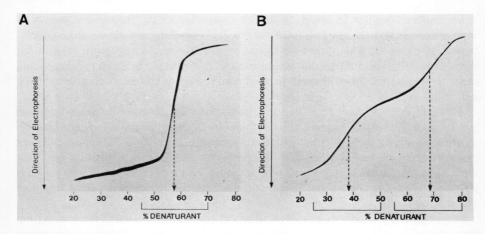

Figure 8. Perpendicular denaturing gradient gels. Both patterns are shown as the negative image of ethidium-stained gels of 450-bp double-stranded DNA fragments. (**A**) A perpendicular denaturing gradient gel of a 450-bp DNA fragment that melts in two domains. The steep transition in mobility at the middle portion of the gel results from the first domain melting at about 60% denaturant, which can be determined by estimating the mid-point of the transition. From this information, one would choose a parallel gel for this DNA fragment with a range of denaturants between 47.5% and 72.5%. (**B**) A perpendicular denaturing gradient gel of a 450-bp DNA fragment that melts in three domains. Two transitions in mobility are seen. The left one, at about 35% denaturant, is due to the first domain melting. The right one occurs at 65% denaturant when the second domain melts. (Reprinted with permission from ref. 30.)

above, the DNA fragment enters the gel as a double-stranded molecule and undergoes a mobility reduction in the middle of the gel, at about 38% denaturant, when the first domain melts. In the second (55−80%) parallel gel, the first melting domain melts immediately as the DNA fragments enter the gel, and this partially-melted molecule is dragged through the gel until the second domain melts, where a larger mobility retardation occurs.

The two DNA fragments shown are typical of most fragments in the size range of 100−1000 bp; generally two or three domains, whose Tms are 10−30% denaturant apart, are seen. For some fragments, two mobility transitions are seen on the perpendicular gel, but their Tms fall fairly close together (1−10% denaturant apart). In these cases, it is best to analyse the fragment on a single parallel gel, with a slightly larger denaturant range (~30%, where the denaturant concentration at the mid-point of the gel is the mid-point between the two Tms).

For perpendicular gel analysis carry out the following steps.

(i)   Digest 15−20 μg of plasmid DNA carrying the target DNA insert with restriction enzymes that excise the insert from the plasmid vector sequences. Remove the enzyme by phenol extraction and precipitate the DNA with ethanol. Re-suspend first in 20−40 μl of TE, then add about 100 μl of neutral loading solution (Section 6.1.2).

Note: it is not necessary to purify the insert (test) fragment away from the vector fragment for the preliminary gradient gels that are used to optimize gel conditions. The sizes of the insert DNA fragments and vector fragments are different enough so that there is a large mobility difference between them. Generally, the vector sequences barely enter the gels due to their large size.

(ii) After pouring a $0-80\%$ perpendicular gel and setting it into the heated aquarium as described in Section 6.1.3, load the sample across the top of the gel with a drawn-out capillary pipette.

(iii) Place the cathode into the upper chamber of the apparatus, and electrophorese at 150 V.

Note: the length of time that perpendicular gels are run is not critical. A rough estimate based on size of the DNA fragment and the % acrylamide used is as follows:

Fragment size (bp)	Acrylamide concentration	Time (h)
50−300	14%	7.5
300−500	7%	5
500−750	7%	7.5
750−1000	7%	10

(iv) After electrophoresis, turn off the power supply, disconnect the cathode and peristaltic pump input and remove the gel plates from the apparatus. Separate the plates and place the gel in a solution of ethidium bromide $(1-5 \ \mu g/ml)$ for about $10-20$ min.

Note: ethidium bromide is a carcinogen. Wear gloves while handling gels soaked in ethidium and dispose of properly.

(v) Visualize the ethidium-stained DNA fragments by UV transillumination. Place a clear ruler (with markings in a colour other than red) across the bottom of the gel, setting zero at the extreme left hand (low denaturant) side of the gel. Obtain a photograph of the gel.

(vi) Use the rulings on the photograph to determine the fractional distance from the 0% denaturant side that a melting transition falls (see *Figure 8*). Use this fractional distance to calculate the percentage denaturant at which the transition occurs.

6.2.2 Determination of electrophoresis times for parallel gels: 'travel schedule' gels

Once the melting profile of a DNA fragment is determined by perpendicular gel electrophoresis and the correct conditions are established for parallel gel analysis, a preliminary experiment to determine the optimal length of time to electrophorese the test fragment in the parallel gel should be performed. The goal of the parallel gel is to electrophorese the DNA fragment for a time that is sufficient to bring the fragment into the denaturant concentration that causes the melting domain in question to melt; in fact, additional electrophoresis time after the domain melts is advisable in order to achieve maximal resolution on the parallel gel.

A simple experiment called a 'travel schedule gel' can be performed to determine the optimal length of time to electrophorese a fragment in parallel gels. Aliquots of the cloned test DNA fragment, excised from the plasmid vector with restriction enzymes, are loaded onto the parallel gel at timed intervals, usually 1 h apart. The gel is electrophoresed and then stained with ethidium as above. By photographing the gel with a ruler as a marker on one side of the gel, it is possible to estimate which of the

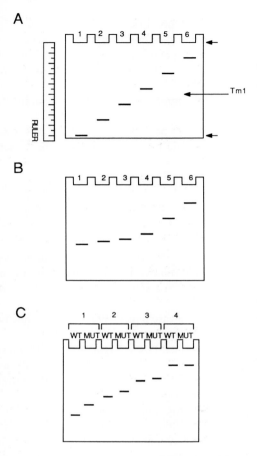

Figure 9. Travel schedule gels. Each drawing depicts the results of a parallel denaturing gradient gel loaded at timed intervals. These 'travel schedule gels' give an indication of the optimal length of time required to electrophorese a DNA fragment for maximum resolution of wild-type and mutant fragments. (**A**) A travel schedule gel showing a DNA fragment that does not undergo a noticeable mobility reduction when it reaches the denaturant concentration at which its first domain melts. The distance between the top and the bottom of the gel is measured with a ruler. This information is used to calculate the position in the gel corresponding to the Tm of the first melting domain (which was determined by a previous perpendicular gel). The timed loading that brings the DNA fragment into this region of the gel is chosen as the minimal time to electrophorese the fragment in subsequent gels. Generally, an additional 1 or 2 h is added to this time for optimal results. (**B**) A travel schedule gel showing a DNA fragment that undergoes a dramatic mobility reduction upon melting of its first domain. Electrophoresis for 12 h (**lane 1**), 10 h (**lane 2**), 8 h (**lane 3**), 6 h (**lane 4**), 4 h (**lane 5**) and 2 h (**lane 6**). By 8 h, the fragment begins to slow down in the gel as melting occurs. The optimal time to run this fragment on subsequent gels is about 9–10 h. (**C**) A travel schedule gel with side-by-side loadings of a wild-type and mutant fragment. If a known mutant fragment is available, a gel of this sort can be used to determine empirically the optimal electrophoresis time that results in maximal separation of the wild-type and mutant fragment.

timed loadings have brought the DNA fragment into the denaturant concentration corresponding to the first Tm (*Figure 9A*).

For many fragments, it is obvious from a glance at the gel which loadings have run far enough to begin melting the fragment because an abrupt mobility retardation of the fragment occurs. In the drawing in *Figure 9B*, one would estimate an optimal electrophoresis time of about 8 h, which is 2 h plus the length of time that the mobility retardation is first noticeable. The degree of mobility retardation that occurs when a domain melts is dependent on a number of factors, including the length of the domain.

Regardless of whether abrupt or only mild retardation occurs, one of the two empirical results obtained in a travel schedule gel can be used to estimate the optimal electrophoresis time for a fragment.

If even one variant of a DNA fragment is available as a cloned DNA molecule, it can be used with the 'wild-type' fragment to optimize the gel conditions for the fragment. In this case, a travel schedule gel is run with side-by-side lanes carrying the wild-type and variant DNA fragments, and gel conditions and electrophoresis times that maximize the resolution between the two fragments are chosen for future gels (see *Figure 9C*).

(i)   Set the parallel gel into the heated aquarium and make the proper connections as described in Section 6.1.4. For travel schedule gels, it is advisable to mark the positions of the lanes of the gel on the glass plates prior to placing the gel in the aquarium.

(ii)   Digest $20-40$ μg of plasmid DNA with restriction enzymes that excise the test insert as described in Section 6.2.1, and re-suspend in TE and neutral loading solution so that the DNA concentration is about 250 μg/ml.

(iii)   Load about $4-5$ μl of sample into a lane toward the left side of the gel. Make a mark with a felt-tipped pen on the gel plate to designate the lane that is loaded to avoid a second loading in the same lane. Electrophorese for 1 h at 150 V.

(iv)   At 1 h intervals, similarly load adjacent lanes with an aliquot of the DNA sample. Although it is difficult to estimate how many loadings will be necessary, we recommend at least six loadings. After the last loading, electrophorese the gel for a length of time estimated to bring the DNA fragment about one-fourth to one-third of the distance from the top of the gel (usually $4-6$ h).

   Note: parallel gels that are designed to detect base changes in a second or third melting domain of a fragment, in which the denaturant concentration is set so that the early domain melts as the DNA fragment enters the gel, generally require longer electrophoresis times than other fragments of similar use. This is because of the decrease in mobility resulting from the melting of the first domain as the DNA enters the top of the gel.

(v)   After the electrophoresis, stain and photograph the gel as described above for perpendicular gels. From the ruled markings and in cases where abrupt mobility reduction occurs, estimate the time of electrophoresis necessary to achieve initial melting, and use this time plus one or two additional hours of electrophoresis for future runs with the fragment.

6.3 Preparation of probes for analysis by DGGE

There are many strategies of probe preparation and sample analysis that can be used to detect base changes in genomic, cloned or PCR-amplified DNA samples by DGGE. Some of these approaches are as follows.

(i)   The use of single-stranded DNA probes, either uniformly labelled or end-labelled, annealed to denatured DNA to produce heteroduplexes with the test DNA samples, followed by direct detection by autoradiography of the gel (17).

(ii)   The use of uniformly labelled single-stranded RNA probes to produce heteroduplexes with the test DNA samples, followed by direct detection by autoradiography of the gel (20).

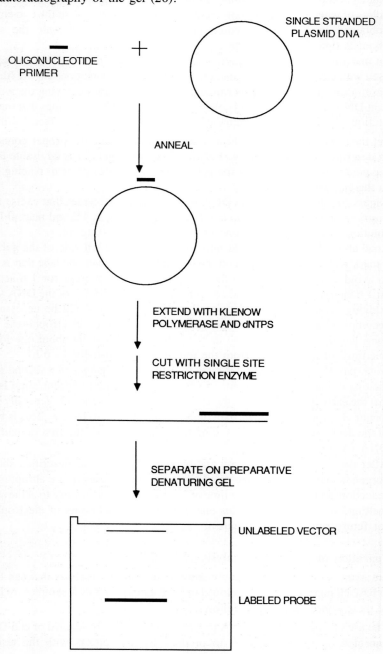

(iii)   The use of double-stranded DNA probes, labelled on only one strand, to anneal the excess unlabelled test DNA sample to form heteroduplexes, followed by direct detection by autoradiography of the gel.

(iv)   The use of uniformly labelled single-stranded RNA probes annealed to mRNA from a cell, followed by direct detection by autoradiography of the gel (see, for example 31).

(v)   The electrophoresis of homoduplexes on a denaturing gradient gel, followed by electroblotting of the gel and hybridizing the blot with a labelled probe.

(vi)   Direct detection of mobility shifts in homoduplexes of cloned or PCR-amplified wild-type and mutant DNA fragments by ethidium bromide staining of the gel.

In all cases, probes should be synthesized that exactly or very nearly correspond to the test DNA fragment generated by PCR or restriction enzyme cleavage. Some degree of 'overhang' of 5' and/or 3' sequence in the probe/test fragment hybrid is acceptable for DGGE: however, one should strive to keep the total length of overhang under 40 nucleotides.

In this section, we describe the procedures used to synthesize two of these types of probes: uniformly-labelled and end-labelled single-stranded DNA probes. To synthesize RNA probes for DGGE, use the procedure exactly as described in Section 5.2 for the RNase cleavage procedure.

6.3.1 *Single-stranded DNA probes*

The procedures that we use for synthesizing both uniformly-labelled and end-labelled single-stranded DNA probes are very similar, and involve the strategy shown in *Figure 10*. The DNA fragment to be tested is cloned into the double-stranded form of a bacteriophage or plasmid vector that produces single-stranded circular DNA molecules of only one sense; the two most widely used systems are the bacteriophage M13 and plasmids that carry the M13 origin of replication. An oligonucleotide primer is annealed to the single-stranded circle at a position on the test insert so that it will prime DNA synthesis, in the 5' to 3' direction, through the test insert. For uniformly labelled probes, one of the dNTPs that is used in the synthesis reaction is labelled with ^{32}P or ^{35}S in the alpha position. For end-labelled probes, the oligonucleotide is labelled at its 5' end with [γ-^{32}P]ATP and cold dNTPs are used in the DNA synthesis reaction. After DNA synthesis, the DNA is digested with a restriction enzyme that cleaves at the end of the test DNA insert, distal to the oligonucleotide annealing site. This reaction produces a linear DNA fragment that is partially double stranded and partially single stranded.

Figure 10. Single-stranded DNA probe synthesis. Single-stranded plasmid DNA is annealed to an oligonucleotide primer, and the primer is extended in the 5' to 3' direction with deoxynucleotide triphosphates and a DNA polymerase. After the polymerization reaction, which generally does not extend the full length of the plasmid, the double-stranded region of the molecule is digested with a restriction enzyme that cleaves at the distal portion of the probe insert fragment. This cleavage produces a linear fragment that is partially double stranded and partially single stranded. The mixture is heated in formamide and run on a preparative denaturing acrylamide ('sequencing') gel to separate the probe strand from the template strand. The degree of separation is large since the two strands are of very different size. The labelled probe strand is then eluted from the gel. Note that the probe can be labelled in two ways with this scheme: (i) with labelled dNTPs in the extension reaction, to produce a uniformly-labelled probe, or (ii) with a 5'-labelled oligonucleotide and cold dNTPs in the extension reaction, to produce an end-labelled probe.

133

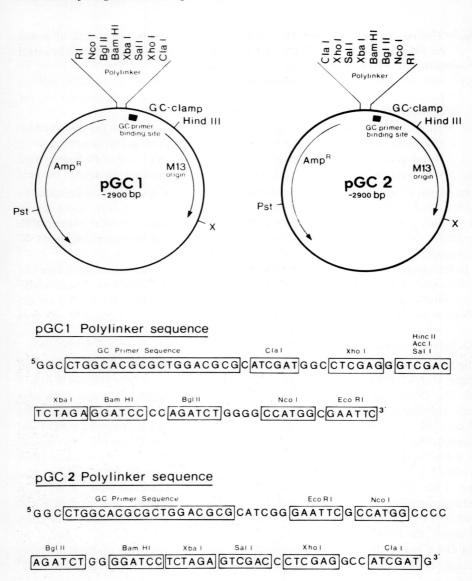

Figure 11. Maps of pGC1 and pGC2. These plasmids carry the M13 origin of replication, which allows the preparation of single-stranded circular forms of the plasmids following infection with M13 helper phage. The only difference between the two plasmids is the orientation of the polylinker cloning region. Thus, both single strands of an insert fragment can be isolated by directionally cloning the fragment into each vector. The arrangement of the origin is such that single-stranded DNA from each vector is primed by oligonucleotides extending 5′ to 3′ in the direction opposite to the arrow representing the M13 origin. The plasmid is composed of the following sequences, counterclockwise from the polylinker site: (**a**) polylinker to 'X'—from nucleotide 4362 of pBR322 decreasing in number to 2440 of pBR322. This segment includes the β-lactamase gene and its control sequences and the pBR322 origin of replication. (**b**) 'X' to *Hind*III—from nucleotide 6000 of M13, decreasing in number to about 5530 of M13. This segment includes the M13 origin of replication isolated by Seed and co-workers (40). (**c**) *Hind*III to polylinker—a 300-bp 'GC-clamp', which is derived from sequences upstream from the human α-globin gene (18,19). (Reprinted with permission from ref. 30.)

By denaturing the DNA and running it on a preparative denaturing polyacrylamide gel, it is possible to purify the shorter single-stranded probe strand from the longer single-stranded template strand.

Three protocols are described below.

(i) *Preparation of single-stranded template DNA*

(a) Clone the test insert DNA fragment into a plasmid vector that carries the M13 origin of replication [examples: pGC1 and pGC2 (*Figure 11*; 30); a series developed by B.Seed (40); Bluescribe M13 (Stratagene, Inc., San Diego, CA)].

(b) Grow a 10 ml culture of an *E.coli* strain carrying the plasmid and an F' until saturated [examples: LE392F' (40); HM101 (41)]. Use LB medium containing the drug for selection of the plasmid DNA (usually ampicillin).

(c) Infect the 10 ml culture with a helper strain of M13 to provide phage components needed for packaging the single-stranded DNA [example: M13rvl (40)]. Use a multiplicity of infection of 20−200. Mix the phage with the culture and leave to incubate at room temperature for 10 min.
Note: a 10 ml culture of *E.coli* grown to $OD_{590} = 0.5$ contains about 4×10^9 bacteria.

(d) Add the 10 ml infected culture to 500 ml of LB medium containing the drug for selection of the plasmid. Shake at 37°C for 10−14 h.

(e) Centrifuge the 500 ml culture for 15 min at 4000 g to pellet the bacteria.

(f) Pour the supernatant into a 1 l centrifuge bottle containing 125 ml of PEG/NaCl solution. Mix well by shaking and leave to stand on ice for 2−12 h. (PEG/NaCl solution = 20% polyethylene glycol 8000 and 2.5 M NaCl).

(g) Pellet the phage/PEG complex by centrifuging at 4000 g for 20 min. Carefully pour off the supernatant.

(h) Suspend the phage pellet in the remaining PEG solution by pipetting. Transfer to a plastic test tube (5−15 ml capacity) and re-pellet the phage by centrifuging at 4000 g for 10 min. Remove all traces of the liquid by aspiration.

(i) Re-suspend the phage pellet in 2 ml of TE buffer. Centrifuge the tube at 4000 g for 15 min to pellet the cell debris. Transfer the supernatant containing the phage in solution into another tube.

(j) Extract the phage solution twice with phenol and once with 2FC (see Section 5.1).

(k) Add sodium acetate to 0.3 M and precipitate the single-stranded DNA with ethanol. Re-suspend the DNA pellet in 0.5−1.0 ml of TE buffer and quantitate by reading the OD_{260}, using an extinction coefficient of 38 $OD_{260}/1$ mg/ml solution.

(ii) *Preparation of uniformly-labelled single-stranded DNA probes.* The materials required are listed below.

(a) Single-stranded DNA template (100−1000 μg/ml).

(b) Oligonucleotide primer (10 pmol/μl). This primer is one of the two that is used to amplify the test DNA fragment by PCR (see Section 4.2). If the nested oligo procedure is used, probe synthesis should be done with one of the two internal oligonucleotides used in the second amplification.

(c)   10 × 3-dNTP mix (dGTP, dCTP and dTTP, each at 1.5 mM).
(d)   Cold dATP solution (1 mM dATP).
(e)   [α-^{32}P]dATP (400 Ci/mmol).
(f)   10 × probe reaction buffer [100 mM Tris−HCl (pH 7.8), 100 mM MgCl$_2$, 500 mM NaCl and 20 mM DTT].
(g)   Large fragment DNA polymerase (Klenow fragment) from *E.coli* (5−20 U/μl).
(h)   The appropriate restriction enzymes.
(i)   Denaturing polyacrylamide gels ('sequencing gels') with spacers and combs for preparation purposes (0.6−1.0 mm thick, with lanes 1.5−2.5 cm wide) and formamide denaturing loading solution (see Section 5.1).

The protocol for probe synthesis generates a single-stranded DNA probe with a specific radioactivity of 10 Ci/mmol in one labelled deoxyribonucleotide (dATP).

(a)   Add the following components to a microcentrifuge tube:
        1 μg of single-stranded template DNA;
        1 μl of oligonucleotide primer (10 pmol/μl).
        2 μl of 10 × probe reaction buffer;
        water to 15 μl.
(b)   Incubate at 45−50°C for 30 min to anneal the oligonucleotide primer to the template.
(c)   Add the following components:
        2 μl of 10 × 3-dNTP mix;
        1 μl of cold dATP solution (1 mM dATP);
        1 μl of [α-^{32}P]dATP (400 Ci/mmol);
        5 U of large fragment DNA polymerase (Klenow).
(d)   Incubate at room temperature for 1 h.
(e)   Heat the reaction mixture to 70°C to stop the DNA polymerization reaction. Add restriction enzyme buffer and water to 100 μl, and then add 5 units of the appropriate restriction enzyme to digest the distal end of the test DNA insert.
(f)   Add 10 μg of carrier tRNA, extract with phenol and precipitate the DNA with ethanol.
(g)   Re-suspend the pellet in 25 μl of formamide loading solution (see Section 5.1).
(h)   Boil the sample for 3−4 min and load immediately into a 1.5−2.5 cm lane of a denaturing polyacrylamide/urea ('sequencing') gel.
        Note: for probes 100−500 nucleotides in length, use a 7% acrylamide gel; for probes 500−1000 nucleotides in length, use a 5% gel.
(i)   Electrophorese at 20−30 W (1000−1500 V) until the bromophenol blue dye runs off the bottom of the gel.
(j)   Locate the labelled probe fragment in the gel by autoradiography. Excise the labelled band from the gel and elute by the crush-soak method (42), using 10 μg of carrier tRNA. After ethanol precipitation, re-suspend the probe pellet in 50 μl of TE buffer.

(iii) *Preparation of end-labelled single-stranded DNA probes.* The method for preparing end-labelled probes is identical to that described above, with the following exceptions.

(a)   The oligonucleotide primer is labelled at its 5′ end with [γ-^{32}P]ATP and polynucleotide kinase as described below.
(b)   A solution of all four dNTPs (unlabelled) is used instead of the 10 × 3-dNTP mix the cold dATP mix, and the [α-^{32}P]dATP solutions in the DNA polymerase reaction.

The following protocol is used to label the oligonucleotide primer with ^{32}P on its 5′ end.

(a)   Mix the following components in a microcentrifuge:
        1 μl of oligonucleotide primer (10 pmol/μl);
        2 μl of 10 × kinase buffer;
        2 μl of DTT (100 mM);
        14 μl of water;
        100 μCi [γ-^{32}P]dATP (1000−7000 Ci/mmol);
        5 units of T4 polynucleotide kinase;
    to a final volume of 20 μl.
        10 × kinase buffer = 0.5 M Tris−HCl, pH 9.5, 0.1 M MgCl$_2$, 10 mM spermidine, and 1 mM EDTA.
(b)   Incubate at 37°C for 1 h.
(c)   Add 10 μg of carrier tRNA, 20 μl of ammonium acetate (7 M) and 80 μl of water. Add 30 μl of ethanol, chill to −70°C for 20 min and centrifuge for 10 min in a microcentrifuge.
(d)   Re-suspend the labelled oligonucleotide in 10 μl of TE. Use 5−10 μl of labelled primer in the probe synthesis reaction.

6.4 Detection of single base changes by denaturing gradient gel electrophoresis

This section describes the approaches used to anneal a labelled probe to test DNA samples and to analyse the resulting hybrids by DGGE. In the case described below, it is not necessary to use a precise ratio of probe to test DNA. In fact, the test DNA sample is in large molar excess to the probe under these conditions.

(i)   Mix the following in a microcentrifuge:
        0.01−0.1 pmol of labelled single-stranded DNA or RNA probe (~1/50 of the total quantity of probe synthesized by the methods above)
        25−100 ng of PCR-amplified test DNA fragment (an amount equivalent to ~1/20 of the total DNA generated from either of the two PCR reactions; see Section 4.2)
        Water to 20 μl
(ii)   Heat the mixture to 95−100°C for 10 min to separate the DNA strands.
(iii) Briefly centrifuge the tube to collect the liquid at the bottom.
(iv)   Add 10 μl of sodium acetate (2.5 M) and incubate the tube at 45°C for 30 min to anneal the probe to the test DNA fragment.
(v)   In the case where a single-stranded circular DNA plasmid that is complementary to the DNA or RNA probe is available, a 'trapping' step that binds up residual unhybridized probe can be performed (15,43). In these cases, after the annealing

reaction, add 1 μg of single-stranded complementary circular DNA to the tube and incubate a further 10 min at 45°C. These trapped molecules remain at the top of the denaturing gradient gel and, even in cases where the test DNA is in excess, this step often results in lowered background. Note that the trapping template is the same single-stranded DNA template used to synthesize the probe in the cases where single-stranded DNA probes are used.

(vi)   Add 70 μl of water and 5 μg of carrier tRNA. Precipitate the nucleic acids with ethanol.

(vii) Dry the pellet and re-suspend in 20 μl of neutral loading solution (see Section 6.1.2).

(viii) Load 10 μl of the sample in a well in a parallel denaturing gradient gel and electrophorese as described in Section 6.2.2.

Note: it is useful to run control lanes with probe alone and probe annealed to PCR-amplified DNA from a cloned template of the same base sequence as the probe. In addition, once a genomic DNA sample has been identified as homozygous wild-type sequence with a given probe, that sample is a useful control to include in subsequent analyses.

(ix)   Dry the gel, expose to X-ray film and analyse the autoradiogram.

7. ACKNOWLEDGEMENTS

Much of the work on RNase cleavage and DGGE was done in the laboratory of Dr Tom Maniatis, and we acknowledge his critical role in the development of these technologies and thank him for his support. We also acknowledge the major role that Dr Leonard Lerman has played in the invention and subsequent development of denaturing gradient gel electrophoresis. We thank Alison Cowie for preparing the figures, Stuart Fischer, Zoia Larin, Brian Seed, Mary Collins, Ezra Abrams, Kary Mullis and Frank McCormick.

8. REFERENCES

1. Flavell,R.A., Kooter,J.M., DeBoer,E., Little,P.F.R. and Williamson,R. (1978) *Cell*, **15**, 25.
2. Geever,R.F., Wilson,L.B., Nallaseth,F.S., Milner,P.F., Bittner,M. and Wilson,J.T. (1981) *Proc. Natl. Acad. Sci. USA*, **78**, 5018.
3. Chang,J.C. and Kan,Y.W. (1982) *New Engl. J. Med.*, **307**, 30.
4. Orkin,S.H., Little,P.F.R., Kazazian,H.H. and Boehm,C.D. (1982) *New Engl. J. Med.*, **307**, 32.
5. Kidd,V.J., Wallace,R.B., Itakura,K. and Woo,S.L.C. (1983) *Nature*, **304**, 230.
6. Pirastu,M., Kan,Y.W., Cao,A., Connor,B.J., Teplitz,R.L. and Wallace,R.B. (1983) *New Engl. J. Med.*, **309**, 284.
7. Botstein,D., White,R., Skolnick,M. and Davis,R. (1980) *Am. J. Hum. Genet.*, **32**, 314.
8. Solomon,E. and Bodmer,W. (1979) *Lancet*, **i**, 923.
9. Kan,Y.W. and Dozy,A.M. (1978) *Proc. Natl. Acad. Sci. USA*, **75**, 5631.
10. Gusella,J.F., Wexler,N.S., Conneally,P.M., Naylor,S.L., Anderson,M.A., Tanzi,R.E., Watkins,P.C., Ottina,K., Wallace,M.R., Sakaguchi,A.Y., Young,A.B., Shoulson,I., Bonilla,E. and Martin,J.B. (1983) *Nature*, **306**, 234.
11. Orkin,S.H., Kazazian,H.H., Antonarakis,S.E., Goff,S.C., Boehm,C.D., Sexton,J.P., Waber,P.G. and Giardina,P.J.V. (1982) *Nature*, **296**, 627.
12. Orkin,S.H. and Kazazian,H.H. (1984) *Annu. Rev. Genet.*, **18**, 131.
13. Gitschier,J., Drayna,D., Tuddenham,E.G.D., White,R.L. and Lawn,R.M. (1985) *Nature*, **314**, 738.
14. Gitschier,J., Wood,W.I., Tuddenham,E.G.D., Shuman,M.A., Goralka,T.M., Chen,E.Y. and Lawn,R.M. (1985) *Nature*, **315**, 427.
15. Myers,R.M. and Maniatis,T. (1986) *Cold Spring Harbor Symp. Quant. Biol.*, **51**, 275.

16. Myers,R.M., Larin,Z. and Maniatis,T. (1985) *Science,* **230**, 1242.
17. Myers,R.M., Lumelsky,N., Lerman,L.S. and Maniatis,T. (1985) *Nature,* **313**, 495.
18. Myers,R.M., Fischer,S.G., Maniatis,T. and Lerman,L.S. (1985) *Nucleic Acids Res.,* **13**, 3111.
19. Myers,R.M., Fischer,S.G., Lerman,L.S. and Maniatis,T. (1985) *Nucleic Acids Res.,* **13**, 3131.
20. Myers,R.M., Maniatis,T. and Lerman,L.S. (1987) In *Methods in Enzymology.* Wu,R. (ed.), Academic Press, New York, Vol. 155, p. 501.
21. Saiki,R.K., Scharf,S., Faloona,F., Mullis,K.B., Horn,G.T., Erlich,H.A. and Arnheim,N. (1985) *Science,* **230**, 1350.
22. Mullis,K., Faloona,F., Scharf,S., Saiki,R., Horn,G. and Erlich,H. (1986) *Cold Spring Harbor Symp. Quant. Biol.,* **51**, 263.
23. Mullis,K.B. and Faloona,F.A. (1987) In *Methods in Enzymology.* Wu,R. (ed.), Academic Press, New York, Vol. 155, p. 335.
24. Maniatis,T., Fritsch,E.F. and Sambrook,J. (1982) *Molecular Cloning, A Laboratory Manual.* Cold Spring Harbor Laboratory Press, Cold Spring Harbor, New York.
25. Freeman,G.J. and Huang,A.S. (1981) *J. Gen. Virol.,* **57**, 103.
26. Winter,E.E., Yamamoto,E., Almoguera,C. and Perucho,M. (1985) *Proc. Natl. Acad. Sci. USA,* **82**, 7575.
27. Fischer,S.G. and Lerman,L.S. (1983) *Proc. Natl. Acad. Sci. USA,* **80**, 1579.
28. Fischer,S.G. and Lerman,L.S. (1979) In *Methods in Enzymology.* Wu,R. (ed.), Academic Press, New York, Vol. 68, p. 183.
29. Saeger,W. (1984) *Principles of Nucleic Acid Structure.* Springer-Verlag, New York.
30. Myers,R.M., Lerman,L.S. and Maniatis,T. (1985) *Science,* **229**, 2423.
31. Smith,F.I., Parvin,J.D. and Palese,P. (1986) *Virology,* **150**, 55.
32. Levinson,B., Janco,R., Phillips,J. and Gitschier,J. (1988) *Nucleic Acids Res.,* **15**, 9797.
33. Wrischnick,L.A., Higuchi,R.G., Stoneking,M., Erlich,H.A., Arnheim,N. and Wilson,A.C. (1987) *Nucleic Acids Res.,* **15**, 529.
34. Wong,C., Dowling,C.E., Saiki,R.K., Higuchi,R.G., Erlich,H.A. and Kazazian,H.H. (1987) *Nature,* **330**, 384.
35. Zinn,K., DiMaio,D. and Maniatis,T. (1983) *Cell,* **34**, 865.
36. Melton,D.A., Krieg,P.A., Rebagliati,M.R., Maniatis,T., Zinn,K. and Green,M.R. (1984) *Nucleic Acids Res.,* **12**, 7035.
37. Krieg,P.A. and Melson,D.A. (1987) In *Methods in Enzymology.* Wu,R. (ed.), Academic Press, New York, Vol. 155, p. 397.
38. Lerman,L.S. and Silverstein,K. (1987) In *Methods in Enzymology.* Wu,R. (ed.), Academic Press, New York, Vol. 155, p. 482.
39. Lerman,L.S., Silverstein,K. and Grinfeld,E. (1986) *Cold Spring Harbor Symp. Quant. Biol.,* **51**, 285.
40. Levinson,A., Silver,D. and Seed,B.J. (1984) *J. Mol. Appl. Genet.,* **2**, 507.
41. Messing,J., Crea,R. and Seeburg,P.H. (1981) *Nucleic Acids Res.,* **9**, 309.
42. Maxam,A. and Gilbert,W. (1980) In *Methods in Enzymology.* Grossman,L. and Moldave,K. (eds), Academic Press, New York, Vol. 65, p. 499.
43. Noll,W. and Collins,M. (1987) *Proc. Natl. Acad. Sci. USA,* **84**, 3339.

CHAPTER 6

The polymerase chain reaction

RANDALL K.SAIKI, ULF B.GYLLENSTEN and HENRY A.ERLICH

1. INTRODUCTION

The polymerase chain reaction (PCR) is an *in vitro* DNA amplification procedure that can, in a matter of hours, isolate and amplify a specific segment of DNA by as much as 10^8-fold (1,2). Such a high degree of target enrichment greatly simplifies any subsequent manipulations of the DNA sample. Some of the applications of PCR include the high-efficiency cloning of genomic sequences (3), the direct sequencing of mitochondrial (4) and genomic DNAs (5−7, see below), the analysis of nucleotide sequence variations (8) and the detection of viral pathogens (9−11).

PCR amplification involves two oligonucleotide primers that flank the DNA segment to be amplified and repeated cycles of heat denaturation of the DNA, annealing of the primers to their complementary sequences and extension of the annealed primers with DNA polymerase (*Figure 1*). These primers are designed to hybridize to opposite strands of the target sequence and are oriented so DNA synthesis by the polymerase proceeds across the region between the primers, effectively doubling the amount of that DNA segment. Since the extension products are also complementary to and capable of binding primers, successive cycles of amplification continue to double the amount of DNA synthesized in the previous cycle. The result is an exponential accumulation of the specific target fragment, approximately 2^n, where n is the number of cycles of amplification performed. Because the primers are physically incorporated into the end of the extension products, they define the primary product of the reaction—a discrete fragment whose length is the distance between the 5′ termini of the primers on the target sequence. A recent technological development involving a thermostable DNA polymerase isolated from the bacterium *Thermus aquaticus* (*Taq*) has substantially improved the performance of the procedure (12). Unlike the thermolabile Klenow fragment of *Escherichia coli* polymerase I used previously, *Taq* polymerase retains its activity after heat denaturation of the DNA and does not need to be replaced during each cycle. In addition to simplifying the reaction, the higher temperature optimum of this enzyme (70−75°C) significantly increases the specificity, yield and length of targets that can be amplified. This chapter will describe PCR amplifications with *Taq* DNA polymerase and strategies for the direct sequencing of PCR-amplified products.

2. BASIC PROCEDURES

2.1 Equipment

The reactions are usually carried out in 0.5 or 1.5 ml Eppendorf-type disposable microcentrifuge tubes. Siliconization of the tubes is optional. The equipment used for

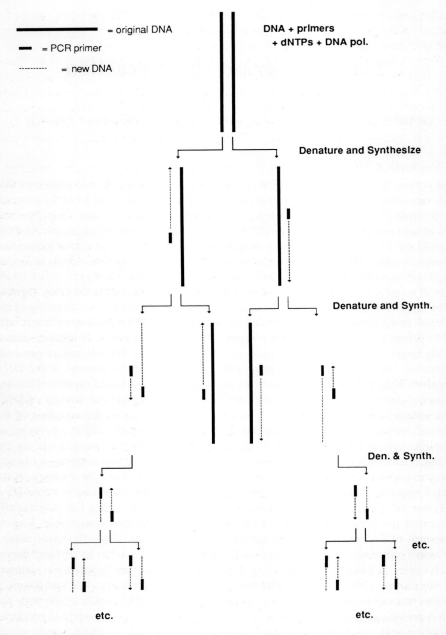

Figure 1. Schematic diagram of the polymerase chain reaction. Only the first two cycles are shown completely. Beginning with the third cycle, the diagram does not show the fate of the original DNA and the extension products made from it. Note that the long primer extensions from the original template can only increase additively with each cycle. In contrast, the short, discrete, primer-terminated copies, which first appear in the third cycle, proceed to double with each subseqeunt cycle and rapidly become the predominant form of amplification product.

heating and cooling may be as simple as water baths set at different temperatures for manual reactions or as sophisticated as a microprocessor-controlled rapid-temperature-changing heat block for fully automated amplifications.

Water baths, oil baths or dry heat blocks are all satisfactory for manual PCR. Water and oil baths provide faster heat transfer than dry heat blocks, but are somewhat messier. If heat blocks are used, the best results are obtained with aluminium blocks that have contoured holes matching the shape of the conical microcentrifuge tubes. Do not use blocks with large diameter holes filled with sand or glass beads. These fillings have poor heat conducting properties and will not maintain their temperature for more than a few cycles.

Because PCR with *Taq* polymerase is done by simply heating and cooling the sample, it is readily amenable to automation and a number of devices have been developed. One approach uses a robot arm to physically transfer a rack containing the samples from water bath to water bath. Another places the samples in a hollow Lucite block and uses a microcomputer or simple timing device connected to washing machine valves to control the flow of hot and cold water into the block. The microprocessor-controlled heat block specifically designed for automated PCR is currently available from Perkin Elmer-Cetus Instruments (PECI). It consists of a 48-sample, electrically-heated, liquid-cooled aluminium block that can be programmed to perform almost any thermal profile.

When properly set up and used, each of these manual and mechanical schemes perform very well, producing amplification products of equivalent quality (i.e. specificity) and yield.

2.2 Reaction components

In addition to the DNA sample to be amplified, a PCR reaction mix includes buffer, deoxyribonucleotide triphosphates, a pair of primers specific for the target sequence and the thermostable *Taq* DNA polymerase. The buffer for PCR with *Taq* polymerase contains 50 mM KCl, 10 mM Tris−HCl, pH 8.4, 2.5 mM $MgCl_2$ and 100 μg/ml gelatin (Difco). Gelatin is recommended over bovine serum albumin because it is less likely to coagulate during the denaturation step and is readily sterilized in an autoclave. Some protocols include 10% dimethyl sulphoxide (DMSO) to reduce the secondary structure of the DNA; our own experience has shown that DMSO can be slightly inhibitory to the polymerase and decrease the yield of amplification product. It is convenient to prepare a 10 × stock solution and store it at −20°C.

Neutralized deoxyribonucleotide triphosphate solutions (dATP, dCTP, dTTP and dGTP) can be obtained from a number of vendors (USB, Sigma, Pharmacia). Considerable savings can be realized by purchasing lyophilized powders and preparing your own aqueous solutions, but they must be neutralized with sodium hydroxide and concentrations accurately determined by UV absorbance before they can be used. Prepare a stock containing 2 mM of each triphosphate and store the resulting 8 mM dNTP solution at −20°C.

PCR primers are typically 20 bases in length. Longer primers may be synthesized but are seldom necessary. Sequences not complementary to the template can be added to the 5' end of the primers. These sequences become incorporated into the double-stranded PCR product and provide a means of introducing restriction sites (3) or regulatory elements (e.g. promoters) at the ends of the amplified target sequence (6). Shorter primers can be used when limited sequence information is available, as long as the problem of primer stability at the extension temperature is taken into account (see below). Store the primers at a concentration of 10 μM in TE buffer (10 mM Tris−HCl, 0.1 mM EDTA, pH 8.0) at −20°C.

The design of effective PCR primers tends to be empirical. Some general guidelines that maximize the probability of obtaining useful primers include the following.

(i)   Selecting primers with an average GC content (~50%) and a random base distribution. Avoid primers with stretches of polypurines, polypyrimidines or other unusual sequences.

(ii)   Avoiding substantial secondary structure. Careful inspection of the sequence of a primer will often reveal potential problems. Computer programs originally designed to fold RNA sequences and dot homology programs are very helpful in uncovering secondary structures.

(iii)   Checking the primers against each other for complementarity. Obviously, primers that anneal to each other will not be able to amplify the target.

Although most primers will work with varying degrees of success, on occasion primers will be synthesized that completely fail to amplify their intended target. The reasons for this remain somewhat obscure. In many of these instances, simply moving the primers by a few bases in either direction will solve the problem.

Taq polymerase may be purchased from a number of vendors (e.g. Perkin Elmer-Cetus Instruments, New England Biolabs, Stratagene). The concentration of enzyme typically used in PCR is 2 units per 100 μl reaction volume. For amplification reactions involving DNA samples with high sequence complexity, such as genomic DNA, there is an optimum concentration of *Taq* polymerase, usually 1−4 units per 100 μl reaction. Increasing the amount of enzyme beyond this level can result in greater production of non-specific PCR products and reduced yield of the desired target fragment (12).

2.3 Thermal cycling parameters

PCR is performed by incubating the samples at three temperatures corresponding to the three steps in a cycle of amplification—denaturation, annealing and extension. Typically, the double-stranded DNA is denatured by briefly heating the sample to 90−95°C, the primers are allowed to anneal to their complementary sequences by briefly cooling to 40−60°C, followed by heating to 70−75°C to extend the annealed primers with *Taq* polymerase. The time of incubation at 70−75°C during primer extension varies according to the length of target being amplified. The ramp time, or time taken to change from one temperature to another, depends on the type of equipment used for heating and cooling (water bath, heat block, etc.). With one exception (described below), this rate of temperature change is not important and, in practice, the fastest ramps attainable are used to shorten the cycle time. However, in order to be certain that the samples reach the intended temperatures, the actual ramp times for a particular set up should be determined by measuring the temperature during a test amplification. A microprobe thermocouple (Cole-Parmer) and digital multimeter are very helpful for this purpose. Typical values for 100 μl reactions in 1.5 ml microcentrifuge tubes with water baths set at 72°, 93° and 55°C are:

(i) $72-93$°C—1 min
(ii) $93-55$°C—1 min
(iii) $55-72$°C—45 sec.

The ramp times for identical samples in heat blocks with contoured holes are usually at least twice these values.

Insufficient heating during the denaturation step is one of the most common causes of failure in a manual PCR reaction. Generally, the temperature of the reaction should reach at least 90°C for strand separation to occur. To provide a margin of assurance, heat the reactions to about 93°C. As soon as the sample reaches 93°C, it can be cooled to the annealing temperature. Extensive denaturation is unnecessary and limited exposure to elevated temperatures helps maintain maximum polymerase activity throughout the reaction. An overlay of about 50 μl of mineral oil may be used to prevent evaporative loss.

The temperature at which annealing is done depends on the length and GC content of the primers. A temperature of $50-60$°C is a good starting point for typical 20-base oligonucleotide primers with about 50% GC content. Because of the very large molar excess of primers present in the reaction mix, hybridization occurs almost instantaneously and long incubation at the annealing temperature is not required. It is sometimes possible to anneal the primers at 72°C, the temperature at which primer extension occurs. In addition to simplifying the procedure to a two-temperature cycle, annealing at 72°C may further improve the specificity of the primer. Unless the primers remain stably annealed, however, the yield of amplification product will be reduced.

In some cases, primers of only $12-15$ bases are available and an annealing temperature of about 40°C is needed. However, primers of that length do not remain annealed at the 72°C extension temperature. The problem can be overcome by using the partial enzymatic activity of the polymerase at lower temperatures to extend the primers by several bases and stabilize them. This is accomplished either by an intermediate incubation at $55-60$°C for a few minutes (manual reactions) or by heating slowly from 40 to 72°C (automated reactions).

Primer extension at 72°C is very near the temperature of maximum activity for the *Taq* DNA polymerase. As mentioned, the incubation time at 72°C depends on the length of the DNA segment being amplified. Allowing 1 min for every 1000 bp of target is usually adequate; even shorter incubation times can be tried. The primer extension step can be eliminated altogether if the target sequence is 150 bases or less. During the thermal transition from annealing to denaturation, the sample will be within the $70-75$°C range for the few seconds required to completely extend the annealed primers.

3. MANUAL AMPLIFICATION OF A GENOMIC TARGET

The protocol described in this section, while intended for the amplification of single-copy genomic DNA targets of up to 500 bp, illustrates the basic principles and techniques of manually performed PCR and can be modified to suit particular applications. The reaction consists of 1 μg of human genomic DNA in a 100 μl volume containing 1 $\times$ Taq

The polymerase chain reaction

Table 1. Reagents for manual PCR amplification of human β-globin.

10 × Taq salts:	500 mM KCl, 100 mM Tris−HCl, pH 8.4, 15 mM MgCl$_2$, 0.1% gelatin
8 mM dNTP;	2 mM dATP, 2 mM dCTP, 2 mM dTTP and 2 mM dGTP, pH 7
PCR primers:	Each 10 μM in TE (10 mM Tris−HCl, pH 8, 0.1 mM EDTA)
	PC03: 5'-ACACAACTGTGTTCACTAGC-3'
	GH21: 5'-GGAAAATAGACCAATAGGCAG-3'

MOLT4 human genomic DNA at 100 μg/ml

Taq DNA polymerase (PECI) at 5000 units/ml

salts, 1 μM each primer, 0.2 mM each dNTPs (dATP, dCTP, dTTP and dGTP— 0.8 mM total) and 2.5 units of *Taq* polymerase. In this particular example, a 250-bp region of the human β-globin gene will be amplified with the PCR primers PC03 and GH21. The reagents are described in *Table 1*.

(i)   Set three water baths at 93, 55 and 72°C. The 93°C bath may need to be covered when not in use to reduce evaporative cooling.

(ii)   To a 1.5 ml microcentrifuge tube add 10 μl of genomic DNA (1 μg), 10 μl of 10 × Taq salts, 10 μl of primer PC03 (100 pmol), 10 μl of primer GH21 (100 pmol), 10 μl of dNTPs (20 nmol each) and 50 μl of water. The final volume is 100 μl.

(iii)   Add 0.5 μl of *Taq* polymerase (2.5 units), mix and overlay with several drops of lightweight mineral oil ($\sim$50 μl).

(iv)   Place the tube in a foam block 'floater'. The soft, polyurethane-type foam works much better than the hard, polystyrene-type which tends to warp in the 93°C water bath.

(v)   Place the block in the 93°C water bath for 2 min. (This is the denaturing step.)

(vi)   Transfer the block to the 55°C bath and leave it for 1 min. (This is the annealing step.)

(vii)   Transfer the block to the 72°C bath and leave it for 1 min. (This is the primer extension step.)

(viii)   Repeat steps (v)−(vii) a total of 30 times, except that subsequent incubations in the 93°C bath should be only 1 min.

(ix)   After the last cycle, leave the sample in the 72°C bath for an additional 4 min to ensure that the PCR product is fully double stranded.

(x)   If desired, the mineral oil cap can be removed by extraction with chloroform. Store the amplified samples at −20°C.

The quality and yield of the reaction can be conveniently assayed by running 5 μl of the sample on a 4% NuSieve agarose gel (FMC) and staining with ethidium bromide. Such a gel is shown in *Figure 2* and contains the results of manual amplification of β-globin gene segments ranging from 110 bp to 1829 bp.

4. DIRECT SEQUENCING OF ENZYMATICALLY AMPLIFIED DNA FRAGMENTS

The analysis of genetic variation in sequences amplified by the PCR has been carried out by sequencing of cloned PCR products (3,8,13) as well as by hybridization with oligonucleotide probes (14). Recently, methods for sequencing amplified DNA without

146

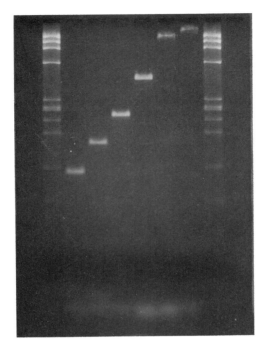

Figure 2. Electrophoretic analysis of manual PCR amplifications of human β-globin gene segments with various primer pairs. The length of the amplified targets, from left to right, are 110, 167, 250, 441, 1232 and 1829 bp. The faint fluorescence at the bottom of the gel is due to excess primer. Flanking molecular weight marker lanes each contain 250 ng of *Hae*III-digested ϕX174-RF (Biolabs). Amplification reactions and subsequent analysis were performed exactly as described in Section 3.

cloning have been developed (4–6). In general, direct sequencing has two major advantages over conventional cloning of PCR fragments into plasmids and viral genomes: (i) it can be more readily standardized (and thus amenable to automation) since it is an *in vitro* system that does not depend on living organisms (bacteria, virus), and (ii) it is faster and more reliable since normally only a single sequence needs to be determined for each sample. By contrast, several cloned PCR sequences need to be determined for each sample in order to distinguish mutations occurring in the original genomic sequence from random misincorporated nucleotides introduced by the DNA polymerase during PCR, and PCR artefacts such as the formation of mosaic alleles ('shuffle clones') by *in vitro* recombination (12).

The ease with which clear and reliable sequences can be obtained directly, without resorting to cloning in bacteria, is determined by (i) the ability of the PCR primers to amplify only the target sequence (usually called the specificity of the primer), and (ii) the method used to obtain a template suitable for sequencing. Although chemical methods for DNA sequencing (Maxam–Gilbert) may be used for direct sequencing of PCR fragments, we will here only consider the more commonly used method of Sanger employing chain terminators. The specificity of PCR primers determines the complexity of the sequences amplified. Ideally, only the target sequence should accumulate exponentially in the reaction. However, many primers do amplify multiple unrelated sequences. By raising the annealing temperature in the PCR reaction this complexity may be reduced (12). Similarly, gel purification of the desired target sequence or use of denaturing gradient gels (15) may be employed to further simplify the complexity of amplified sequences. Finally, if the complexity cannot be reduced further by any of these methods, internal primers may be used in the sequencing reaction to prime only the target sequence.

The second problem associated with direct sequencing of PCR products derives from the ability of the two strands of the amplified fragment to rapidly reassociate, preventing the sequencing primer from annealing to its complementary sequence or blocking the primer—template complex from extending. To reduce this problem either a variant of the standard method for sequencing double-stranded DNA may be employed or single-stranded templates may be produced in the PCR.

4.1 Sequencing of double-stranded templates

Two different protocols are available for preparing the template for sequencing.

(i)   Denature the template with NaOH, transfer to ice and neutralize the reaction, then quickly precipitate the DNA. Resuspend the DNA in buffer and sequencing primer at the desired annealing temperature.
(ii)   Denature the template by heat (95°C) then quickly chill the tube by putting it in a dry ice—ethanol bath to slow down the reassociation of strands. Add the sequencing primer and bring the reaction to the proper annealing temperature.

Both these protocols have been developed to sequence covalently closed double-stranded plasmid templates. Sequencing of PCR products with these protocols is usually more troublesome since the short linear templates are more prone to reassociate than the two strands of a circular plasmid. Detailed protocols for the sequencing of double-stranded PCR products can be found in references 4 and 16.

4.2 Sequencing of single-stranded templates

Sequencing problems derived from template strand reassociation can be avoided by preparing ssDNA templates either by strand separation gels or by generation of single-stranded DNA by the PCR. While agarose strand separating gels may be successfully employed to obtain ssDNA of fragments of more than about 500 bp, they are not well suited for shorter fragments. However, we have recently developed a method that generates the ssDNA of choice in the PCR reaction (*Figure 3*) (6). In this procedure, an asymmetric ratio of the two amplification primers is used to generate dsDNA for the first 20−25 cycles and, when the limiting primer is exhausted, ssDNA for the next 5−10 cycles.

Figure 4A shows the accumulation of dsDNA and ssDNA during a typical amplification of a genomic sequence, using an initial ratio of 50 pmol of one primer to 0.5 pmol of the other primer, in a 100 μl PCR reaction. As expected, the amount of dsDNA accumulates exponentially to the point where one primer is almost exhausted, and thereafter only very slowly. The ssDNA generation appears to start at about cycle 25, the point where the limiting primer is almost depleted. After a short initial phase of rapid increase, the ssDNA accumulates linearly as expected when only one primer is

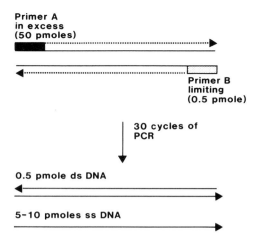

Figure 3. The procedure for generating single-stranded DNA by the polymerase chain reaction. In this example, 50 pmol of one primer and 0.5 pmol of the other is added to the reaction. After about 0.5 pmol of double-stranded DNA has been generated, the limiting primer is essentially used up and single-stranded DNA will start to accumulate at a rate of about 0.5 pmol per cycle of amplification. The resulting single-stranded DNA can be sequenced either by adding more of the limiting amplification primer, or by using an internal primer.

present. A variety of asymmetric primer ratios will yield ssDNA in this way. In *Figure 4B* three different asymmetric ratios 50:5, 50:0.5, 50:0.05 can be seen to yield amounts of ssDNA exceeding that of dsDNA produced after 30 cycles of PCR. In general, a ratio of 50:0.5 will result in about $1-5$ pmol of ssDNA after 30 cycles of PCR. The ssDNA generated can then be sequenced using either the PCR primer that is limiting or an internal primer and applying conventional protocols for incorporation sequencing or labelled primer sequencing. The population of ssDNA strands produced should have discrete 5' ends but may be truncated at various points close to the 3' end due to premature termination of extension. However, for some sequencing primers (e.g. the primer limiting in the PCR), only full length ssDNA can be recruited as template.

A detailed protocol for generation of ssDNA and preparation of sequencing reactions using the modified T7 DNA polymerase (Sequenase, US Biochemicals) is given below.

(i)   Perform the PCR reaction as described elsewhere except for the primer amounts which are set to 50 pmol and 0.5 pmol for a 100 μl reaction. Continue the PCR reaction for $30-35$ cycles. If both strands have to be sequenced also prepare one reaction with the reversed primer ratio.

(ii) After the PCR is finished mix the 100 μl with 2 ml of distilled water and apply to the microconcentrator Centricon 30 (Amicon) and spin at 5000 r.p.m. in a fixed angle rotor to remove excess dNTPs and buffer components.

(iii) Dry down 10 μl of the 40 μl retenate and resuspend in 10 μl of sequencing buffer (40 mM Tris$-$HCl, pH 7.5, 20 mM MgCl$_2$, 50 mM NaCl) containing 1 pmol of sequencing primer (either the limiting primer in the PCR reaction or an internal primer complementary to the ssDNA generated).

(iv) Make the primer$-$template mix 65°C for 2 min and then allow it to cool down to 30°C over a period of 20 min.

(v)   Then add to the mixture:
        1 µl of 100 mM DDT
        2 µl of 1/100 dilution of labelling mix (containing 7.5 µM of each dNTP except
        dATP)
        0.5 µl of [^{35}S]dATP (1000 Ci/mmol), 10 µCi/µl
        2 µl of T7 DNA polymerase (diluted 1/8 in TE)
        Leave the mixture for 5 min at room temperature.
(vi) Add 3.5 µl of the mixture to each of the four tubes with 2.5 µl of termination
        mix (each with 80 µM of each dNTP to 8 µM of the appropriate ddNTP), and
        incubate the reaction at 37°C for 5 min.
(vii) Stop the reaction by adding 4 µl of 95% formamide, 20 mM EDTA, heated to
        75°C for 2 min and load on the sequencing gel.

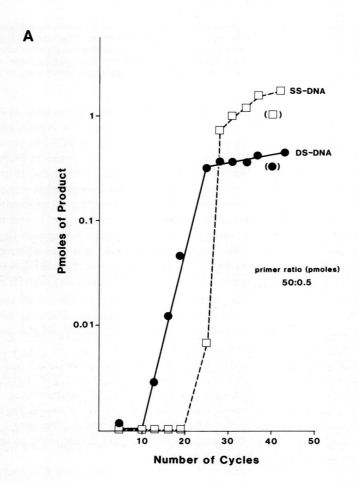

B

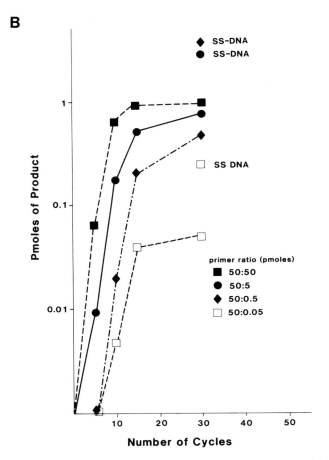

Figure 4. Accumulation of double- and single-stranded DNA of a 242-bp fragment of the HLA-DQα locus at different primer ratios. (**A**) Primer ratio of 50 pmol to 0.5 pmol, up to 43 cycles of amplification. (**B**) Four different primer ratios: 50 pmol:50 pmol, 50 pmol:5 pmol, 50 pmol:0.5 pmol, 50 pmol:0.05 pmol. The amount of single-stranded DNA is given after 30 cycles of amplification only. The curves are based on densitometry scanning of autoradiograms.

5. SUMMARY

The PCR provides a method for synthesizing millions of copies of a specific DNA sequence, significantly facilitating subsequent analysis. Since the amplification reaction generates a specific discrete fragment, cloning as well as direct sequence analysis of the product can be carried out readily. Both DNA and RNA, following reverse transcription into cDNA, can serve as PCR templates and a single DNA molecule can be amplified and detected. With the automation of PCR and the availability of non-radioactive probes, PCR amplification promises to play a major role in the diagnosis of genetic and infectious diseases as well as become a standard procedure in the repertoire of molecular biology techniques.

6. REFERENCES

1. Mullis,K.B. and Faloona,F.A. (1987) In *Methods in Enzymology*. Wu,R. (ed.), Academic Press, New York, Vol. 155, p. 335.
2. Saiki,R.K., Scharf,S., Faloona,F., Mullis,K.B., Horn,G.T., Erlich,H.A. and Arnheim,N. (1985) *Science*, **230**, 1350.
3. Scharf,S.J., Horn,G.T. and Erlich,H.A. (1986) *Science*, **233**, 1076.
4. Wrishchnik,L.A., Higuchi,R.G., Stoneking,M., Erlich,H.A., Arnheim,N. and Wilson,A.C. (1987) *Nucleic Acids Res.*, **15**, 529.
5. Wong,C., Dowling,C.E., Saiki,R.K., Higuchi,R.G., Erlich,H.A. and Kazazian,H.H. (1987) *Nature*, **330**, 384.
6. Stoflet,E.S., Koeberl,D.D., Sarkar,G. and Sommer,S.S. (1988) *Science*, **239**, 491.
7. Gyllensten,U.B. and Erlich,H.A. (1988) *Proc. Natl. Acad. Sci. USA*, in press.
8. Horn,G.T., Bugawan,T.L., Long,C.M. and Erlich,H.A. (1988) *Proc. Natl. Acad. Sci. USA*, **35**, 6012.
9. Kwok,S., Mack,D.H., Mullis,K.B., Poiesz,B., Ehrlich,G., Blair,D., Friedmen-Kien,A. and Sninsky,J.J. (1987) *J. Virol.*, **61**, 1690.
10. Ou,C.-Y., Mitchell,S.W., Krebs,J., Feorino,P., Warfield,D., Kwok,S., Mack,D.H., Sninsky,J.J. and Schochetman,G. (1987) *Science*, **239**, 295.
11. Shibata,D.K., Arnheim,N. and Martin,J.W. (1988) *J. Exp. Med.*, **167**, 225.
12. Saiki,R.K., Gelfand,D.H., Stoffel,S., Scharf,S.J., Higuchi,R., Horn,G.T., Mullis,K.B. and Erlich,H.A. (1988) *Science*, **239**, 487.
13. Scharf,S.J., Friedmann,A., Bautbar,C., Szafer,F., Steinman,L., Horn,G., Gyllensten,U. and Erlich,H.A. (1988) *Proc. Natl. Acad. Sci. USA*, **85**, 3504.
14. Saiki,R.K., Bugawan,T.L., Horn,G.T., Mullis,K.B. and Erlich,H.A. (1986) *Nature*, **324**, 163.
15. Fisher,S.G. and Lerman,L.S. (1983) *Proc. Natl. Acad. Sci. USA*, **80**, 1579.
16. Higuchi,R.G., von Beroldingen,C.H., Sensabaugh,G.F. and Erlich,H.A. (1988) *Nature*, **332**, 543.

CHAPTER 7

DNA fingerprinting

RICHARD A.WELLS

1. INTRODUCTION

In the past few years several sequences which share the property of highly polymorphic length have been discovered in the human genome (1 – 9). These hypervariable regions (HVRs), to which no function has yet been attributed, comprise arrays of short, usually GC-rich, units which are repeated in tandem. It is in this tandemly repetitive structure that the molecular basis of the variability of these sequences lies. The long homologous stretches of repetitive sequence are prone to recombination, presumably through unequal exchange at meiosis or mitosis, or through slippage during DNA replication (10). These recombination events result in allelic differences in the number of repeated units present at an HVR locus and, hence, in length polymorphism (11). Because of this tendency toward variability, the degree of heterozygosity at HVR loci is high (up to 99% for some HVRs) and thus these loci provide highly informative markers for linkage analysis (11 – 13).

Evidence has emerged recently which suggests that HVRs exist as families, the members of which are related by homology of the core unit of their tandem repeats and are scattered throughout the genome (14,15). Cloned HVR loci have been used to generate probes which, when hybridized to genomic DNA under conditions of low stringency, detect alleles from multiple loci related in this way. The resultant complex pattern of hypervariable bands comprises an individual-specific DNA fingerprint (14), the component alleles of which are stable somatically and in the germline, and which are inherited in a Mendelian fashion. The inheritance of these many alleles can thus be followed simultaneously, using a single probe. The evolutionary instability which gives these sequences their hypervariability is not so great as to perturb segregation analysis, and thus DNA fingerprinting may be employed in linkage studies (16). The individual-specific nature of DNA fingerprints also lends itself to exploitation, in forensic science and as a tool by which pedigree structure may be verified with a high degree of certainty.

2. HYBRIDIZATION PROBES FOR DNA FINGERPRINTING

2.1 Minisatellite probes

The first demonstration that a probe based on an HVR core sequence could be used to detect simultaneously many distinct HVR loci was given by Jeffreys and co-workers (14). Using a probe synthesized by end-to-end ligation of the 33-bp repeated unit of the HVR of the human myoglobin gene they detected a pattern of many bands on Southern blots of human genomic digests, several of which appeared to be polymorphic. To isolate these related loci they selected clones from a genomic library, using the

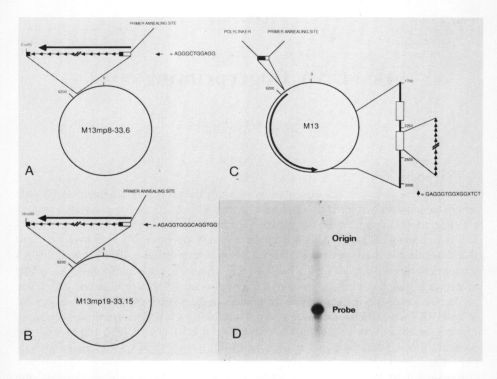

Figure 1. Structure and synthesis of probes for DNA fingerprinting. The small arrows represent the tandemly repeated units of the hypervariable sequence, and the large arrows show the direction of synthesis of the radioactive complementary strand generated in the protocols outlined below. (**D**) is an autoradiograph of a low melting temperature agarose gel through which minisatellite probe 33.6 has been purified by electrophoresis (see *Table 2*). The origin of the track and the probe band are shown.

concatenated myoglobin repeat probe at low stringency. When these clones were, in turn, used as probes, several detected highly complex patterns of hypervariable bands in genomic DNA. These clones contained sequences which shared the tandem repetitive structure of the myoglobin HVR and, taken as a group, shared also a common consensus or core repeat sequence unit. Sequences belonging to this family are known as minisatellites. Two members of the minisatellite family, whose repeated units are distinct variants of the core sequence and which detect distinct sets of hypervariable fragments, are available for research use. These probes, named 33.6 and 33.15, exist as recombinant forms of vectors derived from bacteriophage M13 (17), (*Figure 1*). They each detect approximately 15 highly polymorphic bands in the size range 4−20 kb. The mean heterozygosity of these bands is approximately 90%, with variability being proportional to fragment length and the largest fragments being virtually 100% heterozygous in the population.

2.1.1 *Preparation of minisatellite fingerprinting probes*

Single-stranded templates for probe synthesis can be rapidly and easily generated using standard protocols (summarized in *Table 1*). Template prepared by the protocol in *Table*

Table 1. Preparation of single-stranded M13 template.

A. *Preparation of competent cells*

1. Inoculate 5 ml of 2 × TY[a] with a single colony of JM101[b]. Grow to stationary phase by incubating with shaking at 37°C overnight.
2. Add 500 μl of overnight JM101 culture to 50 ml of 2 × TY. Incubate with shaking at 37°C until the OD_{550} of the culture is ~0.3.
3. Chill 30 ml of 50 mM $CaCl_2$ on ice.
4. Pellet the cells by centrifugation at 3000 g for 5 min. Discard the supernatant.
5. Re-suspend the cells gently in 25 ml of chilled 50 mM $CaCl_2$. Let stand on ice for 20 min.
6. Pellet the cells by centrifugation at 3000 g for 5 min. Discard the supernatant.
7. Re-suspend the cells gently in 5 ml of $CaCl_2$. Let stand on ice for 2 h.

B. *Transformation of competent cells with single-stranded M13 DNA*

1. Dilute single-stranded M13 DNA to 1 ng/μl in TE. Mix thoroughly.
2. Pipette 400 μl aliquots of competent JM101 cells into four sterile 15 ml tubes. Add 1, 5, 10 or 50 μl of the diluted single-stranded DNA solution to each. Mix by gently flicking the tubes and let stand on ice for 1 h.
3. Transfer the tubes to a 42°C water bath for 3 min, then place immediately back on ice.
4. To each tube, add 40 μl of XGAL (2% in dimethylformamide), 40 μl of 100 mM IPTG and 200 μl of JM101 from the overnight culture. Add 3 ml of molten top agar[c] (kept at 40−50°C in a water bath) and mix by inversion. Pour immediately onto an 'H-agar'[d] plate. Allow the top agar overlay to solidify for 5 min, then invert the plates and transfer to a 37°C incubator. After overnight incubation, plaques will be evident—blue plaques represent M13 clones containing no insert, white plaques are recombinants.
5. While the plates are incubating, prepare a fresh 5 ml overnight culture of JM101 in 2 × TY.

C. *Preparation of single-stranded template*

1. Dilute the fresh JM101 culture 1:100 in 2 × TY.
2. Dispense 1.5 ml of diluted culture into a sterile 15 ml tube for each plaque which is to be processed.
3. Inoculate each 1.5 ml culture with phage by stabbing an individual plaque with a sterile plastic microloop and then swirling the loop in the culture.
4. Incubate the cultures at 37°C with shaking for 5 h.
5. Transfer the cultures to microcentrifuge tubes and pellet by spinning in a microcentrifuge for 5 min. Transfer the supernatants to fresh microcentrifuge tubes containing 200 μl of PEG/NaCl[e]. Mix by inversion, then leave to stand at room temperature for 15 min.
6. Centrifuge for 5 min. Pour off the supernatant and centrifuge for 2 min. Remove the remainder of the supernatant with a drawn-out Pasteur pipette. Small PEG pellets should be visible at this stage.
7. To each tube add 100 μl of TE and 50 μl of Tris-saturated phenol. Vortex for 15 sec and let stand for 10 min. Vortex again for 10 sec and centrifuge for 5 min.
8. Transfer each aqueous (upper) phase to a fresh tube. Add to each 10 μl of 3.0 M sodium acetate pH 5.5 and 300 μl of cold (−20°C) absolute ethanol. Vortex briefly and precipitate overnight at −20°C.
9. Pellet DNA by centrifuging for 15 min. Pour off the supernatants and wash the pellets with 500 μl of cold absolute ethanol. Remove the supernatants with a drawn-out Pasteur pipette and dry the pellets under vacuum for 3 min. Re-suspend each pellet in 50 μl of TE.
10. To assess the concentration and molecular weight of the harvested DNA, run a 2 μl aliquot of each sample, along with aliquots of 100 ng of the DNA used to transform the cells and 100 ng of M13 vector DNA.

[a]2 × TY is, per litre: 16 g of bacto tryptone, 10 g of yeast extract, 5 g of NaCl.
[b]*Escherichia coli* JM101 is: *lacpro, thi, supE*, F'traD36, *por*AB, *lac* 1[q]Δ Z M15: and must be maintained on glucose/minimal medium to select for the F' plasmid.
[c]Top agar is, per litre: 10 g of bacto tryptone, 8 g of NaCl, 8 g of agar.
[d]H agar is, per litre: 10 g of bacto tryptone, 8 g of NaCl, 12 g of agar.
[e]PEG/NaCl is: 20% polyethylene glycol 6000, 2.5 M NaCl (store at 4°C).

1 will ordinarily be at a concentration of approximately 100 ng/μl, but it is best to estimate the concentration of a given batch by comparison with standard solutions on an ethidium bromide-stained agarose gel.

Since the template molecule is single-stranded and circular it cannot be labelled by random priming or nick translation methods, and a primer extension technique must

Table 2. Preparation of minisatellite probes.

Radiolabelling probes 33.6 and 33.15

1.   To a microcentrifuge tube add 400 ng of single-stranded template DNA, 4 ng of 17-mer sequencing primer, 1 μl of 10 × Klenow buffer and water to a final volume of 10 μl. Incubate at 55−60°C for at least 1 h. (The annealed template can be stored at −20°C at this stage.)
2.   Spin the tube briefly in a microcentrifuge. Add 10 μl of AGT mixture, 6 μl of TE buffer and 30 μCi of [α-^{32}P]dCTP. Add 6 units of DNA polymerase I (Klenow fragment) and mix by pipetting gently up and down. Incubate at 37°C for 45 min.
3.   Add 2.5 μl of 0.5 mM dCTP. Continue incubation at 37°C for 15 min.
4.   Add 3 μl of *Eco/Hin*d 10 × restriction buffer, 3 μl of 10 mM spermidine trichloride pH 7 and 15 U of the appropriate restriction enzyme (*Eco*RI for 33.6, *Hin*dIII for 33.15). Mix by pipetting up and down. Incubate at 37°C for at least 1 h.
5.   Add 5.2 μl of alkali stop solution. Mix by flicking the tube and allow to stand for 5 min.
6.   Add 5 μl of agarose gel loading buffer.

Electrophoretic purification of probe

1.   Prepare a 1.2% low melting temperature agarose gel in Tris−acetate−EDTA buffer (it is convenient to do this during the incubation following step 4 of the labelling procedure).
2.   Load and run the gel in a cold room at 7 V/cm for 2 h. The bromophenol blue dye front should have migrated ~6 cm from the origin.
3.   The gel is very radioactive. Wearing gloves and goggles, slide the gel out of the gel tray and onto a sheet of clingfilm in an X-ray cassette. In a dark room place a sheet of pre-flashed X-ray film over the gel. Develop the film after a 5 min exposure.
4.   The autoradiograph should show a very strong probe band and a weak background fog against which the gel slot should be visible, see *Figure 1D*. Measure the distances from the well to the leading and trailing edge of the probe band and cut the corresponding chunk from the gel using a scalpel. The probe migrates at about the same position as the bromophenol blue dye front, while any unincorporated nucleotide gives a more diffuse band further down the gel.
5.   Transfer the agarose chunk to a 5 ml plastic tube. Add 500 μl of sterile distilled water and 10 μl of competitor DNA solution. Melt the probe at 100°C for 5 min.
6.   To measure the specific activity of the probe take a 10 μl aliquot and measure the radioactivity by scintillation counting. 400 ng of template should yield $1-3 \times 10^7$ c.p.m. total.

Preparation of competitor DNA solution

1.   Prepare a 10 ml solution of human genomic DNA in TE buffer at a concentration of 0.5 μg/μl.
2.   Shear by sonicating at 50 W for ten 15 sec bursts separated by 5 sec intervals.
3.   Check the fragment size by running a 1 μl aliquot on a small agarose gel. The average fragment size should be 200−600 bp.

Stock solutions for probe synthesis

TE buffer: 10 mM Tris pH 8.0, 1 mM EDTA
dNTP stocks: 0.5 mM dATP, dGTP, dTTP, dCTP pH 7.0
AGT mix: equal volumes of TE buffer and 0.5 mM dATP, dGTP, dTTP
10 × Klenow reaction buffer: 100 mM Tris pH 8.0, 50 mM MgCl$_2$
10 × *Eco*RI/*Hin*dIII reaction buffer: 100 mM Tris, 600 mM NaCl, 70 mM MgCl$_2$,
   70 mM β-mercaptoethanol
Alkali stop solution: 1.5 M NaOH, 0.1 M EDTA

be employed (*Table 2*, *Figure 1*). In this technique, a synthetic oligonucleotide complementary to a sequence 3′ to the minisatellite insert is allowed to anneal to the template in solution. This provides a 3′-hydroxyl group onto which DNA polymerase I (Klenow fragment) can catalyse the addition of labelled deoxyribonucleotides. After a suitable reaction time, when polymerization has occurred across the entire length of the insert, a restriction enzyme can be used to cleave the new partially double-stranded molecule on the 5′ side of the insert. If the resultant molecule is broken into separate single strands by treatment with alkali, then the radiolabelled strand complementary to the minisatellite sequence can be purified by electrophoresis through a low melting temperature agarose gel, located by autoradiography and isolated by cutting the hot

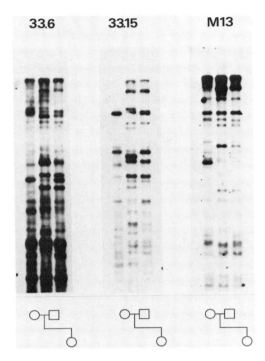

Figure 2. DNA fingerprints generated by sequential re-probing of a single filter. 5 μg of DNA from a woman and her parents was digested with *Mbo*I and electrophoresed through a 1.0% agarose gel at 2 V/cm for 48 h, then blotted onto a nylon membrane. This membrane was then hybridized, in turn, to 33.6, 33.15 and M13 probes, and autoradiographed. The probes were eluted between hybridizations using the protocol described in Section 4.4. A high frequency of band sharing is observed between the paternal and maternal fingerprints because the parents are second cousins.

band from the gel (*Figure 1*). The DNA thus isolated is labelled to high specific activity ($\sim 10^8$ c.p.m./μg) and is relatively free of unincorporated radionucleotide.

2.2 M13

Recently, Vassart *et al.* (18) reported the astonishing finding that a sequence from the genome of the single-stranded filamentous bacteriophage M13 detects a family of hypervariable sequences when used as a hybridization probe against human genomic DNA. The sequence is part of the phage gene III, which encodes a protein involved in attachment to the bacterial F pilus in the process of infection, and comprises a pair of GC-rich tandem repeats (*Figure 1*). These sequences are swamped by the eukaryotic competitor DNA contained in most hybridization buffers, and so it is necessary to omit this component when this sequence is used as a probe to generate a DNA fingerprint. The family of HVRs detected by the M13 gene III tandem repeat is distinct from those detected by the minisatellite probes 33.6 and 33.15 (*Figure 2*), but shares with them the property of a high degree of polymorphism, with allele frequencies diminishing as allele size increases. The mean frequency of band sharing between unrelated individuals of European descent is approximately 0.20 for bands larger than 2 kb of which 15−20 can be detected per individual (R.Wells, unpublished observations). The

DNA fingerprinting

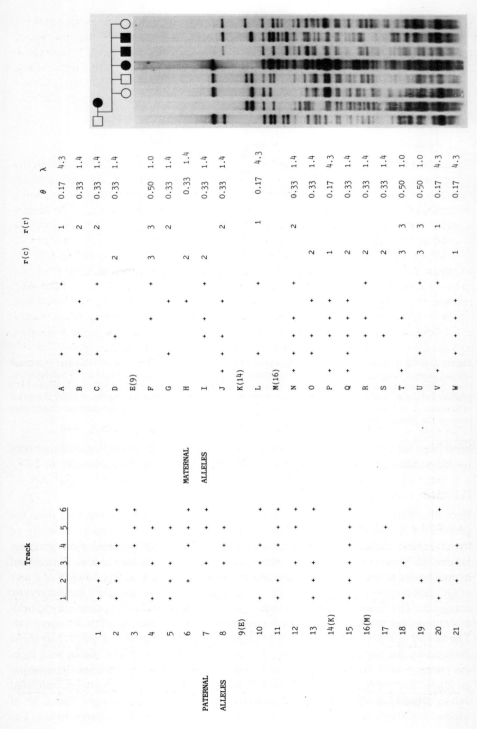

158

M13 tandem repeat thus detects a DNA fingerprint of a complexity comparable to those which are seen when the minisatellites are employed, and allows the examination of an additional 15−20 hypervariable loci in human and other mammalian DNA. Because the M13 repeat is free of recognition sites for the restriction enzymes *Hin*fI, *Mbo*I, *Hae*III and *Alu*I, and because the size distribution of informative bands detected is similar to that of the minisatellite fingerprints, it is convenient and efficient to prepare Southern blots for hybridization to these probes in turn (see *Figure 2*).

2.2.1 *Preparation of M13 gene III fingerprinting probe*

The tandem repeats of M13 are found at positions 1700−1900 and 2300−2500 in the phage genome. It is possible to use this information and the properties of the M13 vector phages to synthesize a hybridization probe of high specific activity in a very simple procedure (*Table 3*). If the synthetic 17-mer sequencing primer is annealed to the single-stranded form of any of the M13mp vectors and Klenow polymerase is used to add labelled deoxynucleotide triphosphates (dNTP's), then the reaction can be stopped by the addition of salt after an incubation period. If the incubation is timed so that the polymerization is halted before the gene III region is crossed, then the new strand of up to 4.5 kb contains the radiolabelled nucleotides and the proximal 2.5 kb of the M13 genome, including the tandem repeats, remains single-stranded. This allows the tandem repeats to hybridize to their filter-bound complements when the labelled phage is used as a hybridization probe. If the free nucleotides are separated from the probe molecules then the specific activity can be measured and is usually in the region of $1-2 \times 10^9$ c.p.m./μg template DNA. Passing the probe through a Sephadex column seems, however, to reduce its efficacy and it is best to avoid this step and use the final salted reaction mixture (step 3, *Table 3*), without further purification, as probe.

3. GEL ELECTROPHORESIS AND FILTER TRANSFER OF DNA FOR FINGERPRINT ANALYSIS

Gel electrophoresis of DNA has been definitively reviewed (19). It is worthwhile, however, to outline briefly the particular conditions which optimize the efficiency of fingerprint analysis.

3.1 **Choice of restriction enzyme**

The molecular basis of the variability of members of the HVR families discussed here consists of the number of repeated 'core' units which exist at a given locus. To expose this sort of variation with the greatest resolution it is necessary to digest the DNA with

Figure 3. Segregation of hypervariable fragments detected by minisatellite probe 33.15. DNA samples (3 μg) were digested with *Hin*fI, separated by electrophoresis through a 0.8% agarose gel at 1.5 V/cm for 72 h and transferred to a nylon membrane as described in the text. The individuals are related as indicated in the family tree, with circles representing females and squares males. The dark symbols show the segregation of an autosomal dominant trait from the mother to three of her children. The segregation of fingerprint bands is scored in the table, with values of the likelihood ratio λ (see Section 5) given for each of the bands scorable in the mother, r(c) and r(r) are the numbers of recombinant offspring observed given the hypothesis of linkage in coupling and repulsion, respectively. The paternal alleles are numbered 1−21 and the maternal alleles are lettered A−W in descending order of size. Alleles marked with an asterisk could not be scored because they co-migrated too closely with a band in the other parent (indicated in brackets).

Table 3. Radiolabelling M13 vector for DNA fingerprinting.

1.	To a microcentrifuge tube add 400 ng of single-stranded M13 vector DNA (e.g. M13 mp8), 4 ng of 17-mer sequencing primer, 1 μl of 10 × Klenow buffer and water to a final volume of 10 μl. Incubate at 55−60°C for at least 1 h.
2.	Spin briefly in a microcentrifuge. Add 10 μl of AGT mixture[a], 6 μl of TE[a] buffer, 30 μCi of [α-^{32}P]dCTP and 2 units of DNA polymerase I (Klenow fragment). Mix by pipetting up and down.
3.	Incubate at 37°C for 15 min, then stop the reaction by adding 70 μl of 3 × SSC.

[a]See *Table 2*.

a restriction enzyme which cuts frequently but not within the repeated unit of the HVR. For the HVRs discussed here, *Hin*fI, *Mbo*I, *Alu*I and *Hae*III are good choices.

Occasionally in a DNA fingerprint two bands will co-migrate so closely as to prohibit their resolution. In such cases it is often helpful to repeat the fingerprinting with another enzyme digest, as this has the effect of 'shuffling' the restriction fragments containing the HVRs and giving another chance to resolve the bands in question. When restriction enzymes recognizing a 4-bp sequence are employed this shuffling effect is ordinarily fairly slight, especially for high molecular weight bands, as the proportion of the length of these bands comprising flanking sequence is small, regardless of which enzyme is chosen. Thus, particular fingerprint bands can still be identified even though their migration in the new fingerprint will have been slightly altered. The restriction enzyme *Hae*III seems to be the most useful in achieving this shuffling effect (R.Sherrington, personal communication).

3.2 Preparation of DNA samples for electrophoresis

DNA obtained by usual extraction procedures (20) is of sufficiently high quality to permit fingerprinting analysis. Best results are obtained if 1−5 μg of DNA is digested per gel track. Because not only the presence but also the intensity of a given fingerprint band is relevant, it is important to achieve evenness of loading across all tracks on fingerprinting gels. Optical density measurements of DNA solutions are notoriously unreliable indices of DNA concentration, and should be regarded only as crude estimates. It is safest to run a small test gel on aliquots of the digested DNA to check for completeness of digestion and consistency of loading, and to adjust on this basis the amount of DNA loaded in each track of the final gel.

Although it is sometimes necessary to ethanol precipitate DNA from the restriction digest to reduce the volume for loading onto the gel, it need not be done routinely to 'clean up' the digest, and good results are obtained when the crude digest is loaded onto the gel. If a sample is precipitated, it is of critical importance that it is washed in 70% ethanol to remove excess salt, which will otherwise perturb migration of DNA fragments, and also that the sample is vacuum dried before it is resuspended in loading buffer, as traces of ethanol cause the sample to float out of the well.

3.3 Electrophoresis conditions

The optimal conditions of electrophoresis for the fingerprinting gel depend upon the information required from the DNA fingerprints. If the data are to be used to verify pedigree structure, or to identify positively a mislabelled sample or to compare samples from different tissues of an individual, then it is necessary to examine as many bands

Table 4. Buffers for electrophoresis.

TAE: 40 mM Tris−HCl, 5 mM sodium acetate, 1 mM EDTA pH 7.7
6 × loading buffer: 0.25% bromophenol, 0.25% xylene cyanol, 15% Ficoll type 400, in water
Gel staining solution: ethidium bromide 50 μg/l in TAE

Table 5. Preparation of fingerprinting gels for transfer.

(NB   Wear gloves while handling gels and blotting hardware.)
1. Place the gel in its tray, in a large plastic box. Add 50 ml of 0.25 M HCl and soak at room temperature on a rocking platform for 15 min. Pour off the HCl and add 500 ml of fresh 0.25 M HCl. Soak for another 15 min. (This partially depurinates the DNA and improves transfer of higher mol. wt fragments.)
2. Pour off the acid solution and rinse the gel with distilled water. Add 500 ml of 0.5 M NaOH/1.5 M NaCl. Soak with rocking for 15 min at room temperature. Pour off the solution and repeat.
3. Pour off the alkali solution and rinse the gel with distilled water. Add 500 ml of 0.5 M Tris pH 7.5/1.5 M NaCl. Pour off and repeat. The gel is now ready for DNA transfer to a nylon membrane.

as it is possible to resolve. In these cases it is most useful to choose a 1.0% agarose gel, run at 2 V/cm for 48 h with two changes of running buffer. This allows resolution of bands in the size range 2−25 kb. For segregation analysis it is more important to be able to resolve the hyperpolymorphic higher molecular weight bands with a high degree of confidence. Here it is better to run a 0.8% gel at 1.5 V/cm for 72 h with five changes of running buffer. This runs all fragments smaller than about 3 kb off the gel, leaving the more highly heterozygous and informative larger bands with optimum separation.

The loading buffer and running buffer (TAE) used for fingerprinting gels are shown in *Table 4*.

3.4 Southern transfer

After electrophoresis the gel is stained in ethidium bromide (*Table 4*) and photographed on a short-wave UV transilluminator to ensure that the samples have migrated far enough and to record the position of the markers.

Southern blotting of the gel is essentially by traditional methods, summarized in *Table 5*.

4. HYBRIDIZATION CONDITIONS AND WASHING OF FILTERS

The most critical factors in generating DNA fingerprints with HVR probes are the conditions for hybridization and washing filters. The optimal conditions differ for the different probes.

4.1 Minisatellites

4.1.1 Pre-hybridization

(i)   Wet the filters in distilled water and place them in a plastic freezer bag, two filters per bag, each with the side to which the DNA is bound facing outwards.
(ii) Add 15 ml of heparin pre-hybridization buffer (21) (*Table 6*) which has been warmed to 37°C.
(iii) Incubate the bags at 37°C for at least 3 h.

Table 6. Buffers for hybridization.

	Formamide	SSC[a]	Dextran sulphate	SDS	Heparin
Pre-hybridization	50%	3×	–	0.2%	50 µg/ml
Hybridization	50%	3×	5%	0.2%	200 µg/ml

[a]SSC is standard saline citrate: 150 mM sodium chloride, 15 mM sodium citrate pH 7.0.

4.1.2 *Hybridization*

(i) Warm 7.5 ml of heparin hybridization buffer to 37°C.

(ii) Add probe ($4 \times 10^6 - 10^7$ c.p.m.) and mix by inversion.

(iii) Cut open the bag containing the pre-hybridization filters and squeeze out the buffer using a roller.

(iv) Add to the bag the hybridization buffer containing the probe and re-seal the bag, being careful not to trap pockets of air.

(v) Spread the probe evenly using a roller and incubate the bag overnight at 37°C.

4.1.3 *Washing*

(i) Warm 1 litre of a solution containing 1.0 × standard saline citrate (SSC) and 0.1% sodium dodecyl sulphate (SDS) to 65°C in a large sandwich box.

(ii) Cut open the bag in a sink and squeeze out the probe solution.

(iii) Place the hot filters in the sandwich box and incubate them at 65°C for 15 min with shaking, then transfer the filters to a box containing fresh washing solution at 65°C.

(iv) After another 15 min incubation at 65°C, transfer the filters from the washing solution onto some paper towels and monitor the radioactivity with a hand-held GM monitor.

(v) Repeat the 15 min 65°C washing incubations until a plateau is reached.

For probes of exceptionally high specific activity it is impossible to wash to low radioactivity levels and shorter exposure times are advised.

4.2 **M13 fingerprinting**

4.2.1 *Pre-hybridization*

As for minisatellites.

4.2.2 *Hybridization*

(i) Warm 10 ml of heparin hybridization buffer to 37°C, add 50 µl of unboiled probe (one half of one reaction) and mix by inversion.

(ii) Squeeze the pre-hybridization buffer out of the bag and add the probe.

(iii) Re-seal the bag, spread the probe by rolling and incubate at 37°C overnight.

4.2.3 *Washing*

(i) Squeeze the probe solution from the bag and transfer the filters to a large sandwich box.

(ii)   Add 1 litre of $3.0 \times$ SSC/0.1% SDS and incubate with shaking at room temperature for 30 min with one change of wash solution.

(iii)   Wash at $2.0 \times$ SSC/0.1% SDS at room temperature for 15 min. After this the filters should be ready for autoradiography.

4.3 Autoradiography

(i)   While the filters are still slightly damp wrap them in clingfilm and mount in an X-ray cassette with intensifying screens.

(ii)   Develop the film after an overnight exposure at 70°C.

(iii)   Occasionally a high degree of background signal is apparent at this stage. If this is the case, re-wash the filters for 45 min at 65°C in $1.0 \times$ SSC/0.1% SDS and repeat the autoradiography.

(iv)   After the overnight 'check' exposure, it is best to maximize the resolution of bands with a $5-10$ day exposure with no intensifying screens.

4.4 Removal of probe from nylon filters

In order for the maximum information to be derived from a filter, it may be probed with all three fingerprinting probes, one after the other, with elution of probe between hybridizations (*Figure 2*). This elution can be accomplished without significant loss of filter-bound DNA or damage to the filter if a nylon membrane is employed (e.g. Hybond-N, Amersham International). Removal of probe bound to a nylon filters is easily achieved by the following method.

(i)   Warm 1 litre of alkali wash solution (0.4 M NaOH) to 45°C.

(ii)   Incubate $1-4$ radioactive filters in a large sandwich box with this solution.

(iii)   After 30 min, remove the filters from the alkali wash solution and place them in 1 litre of neutralizing solution (0.2 M Tris−HCl pH 7.5, $0.1 \times$ SSC, 0.1% SDS) in another sandwich box.

(iv)   Incubate at 45°C for a total of 30 min, with a change of neutralizing solution after 15 min.

After a brief rinse in distilled water, the filters are ready to be re-used. Using this elution protocol, and with ordinary precautions in storage of filters between usages, good results can be obtained for as many as $8-10$ hybridizations. Inevitably, however, some filter-bound DNA is lost each time a blot is stripped of probe. Of these probes, 33.15 seems to bind the most avidly and to require the least DNA to produce a useful result. It should therefore be reserved until last if these probes are to be used several times against a single filter.

5. THE USE OF DNA FINGERPRINTING IN LINKAGE ANALYSIS

5.1 Rationale

The utility of the 'reverse genetic' approach to the study of inherited disease has become widely recognized. Recently our understanding of the biology of Duchenne muscular dystrophy, cystic fibrosis and retinoblastoma has been enhanced by research efforts which have begun with examination of the inheritance of the conditions and thence moved to the localization of the genes responsible and the molecular characterization of those genes (22−24). Although a large number of disease genes have been analysed in this fashion, many remain to be mapped. Occasionally the chromosomal localization of a gene of interest can be deduced readily if individuals who suffer an aberrant phenotype through alteration of the expression of that gene are observed consistently to harbour abnormalities of a specific chromosomal region (for example the 13q14 deletions in retinoblastoma). For most inherited illness, this fortuitous circumstance cannot be exploited and genetic linkage analysis must be undertaken if the gene is to be localized.

The application of restriction fragment length polymorphisms (RFLPs) to linkage analysis has become commonplace, and has greatly enhanced the rapidity with which genes can be isolated (25). There are, however, difficulties inherent in this approach. Most RFLPs are merely dimorphisms and therefore can have no more than 50% heterozygosity. In order for a given mating to provide any linkage information at least one of the parents must be heterozygous at both trait and marker loci. Thus, a high proportion of families subjected to an RFLP linkage analysis will be entirely uninformative. This is a very inefficient use of pedigree resources.

As a tool in linkage analysis, DNA fingerprinting finds its application in the cases of diseases for which no clue to the nature of the genetic pathology exists (i.e. for conditions not involving a known biochemical defect, or in which neither cytogenetic nor physiological data suggest any candidate locus) or for which linkage to candidate loci has been excluded. As there is no easy way of determining which fingerprint bands in one individual represent the locus of a band in another individual, the usual practice of obtaining linkage data from a number of small pedigrees is impossible. The complexity of DNA fingerprints makes it impossible to follow reliably the inheritance of allele pairs, but within an extended large pedigree and certainly within a sibship a single band of a given size can, for the sake of initial analysis, be assumed to represent a single locus. For these reasons, a large single pedigree is an absolute requirement (and a very large sibship a preference) for traditional segregation analysis of DNA fingerprint bands. This sort of pedigree resource is rarely available for disorders inherited in a recessive fashion, but frequently is for dominant conditions.

Because of the simplifying assumption made regarding the allelism of a fingerprint band, classical likelihood methods cannot be applied to linkage analysis of data regarding the segregation of fingerprint bands. It is therefore inappropriate to talk about lod scores in this context, but permissible to estimate the odds of linkage. This should therefore be regarded as a rapid preliminary screening method for linkage. If a promising preliminary result turns up, that is if a fingerprint band is found which segregates with the disease phenotype in a pedigree, then that specific band must be isolated and cloned (26). In this way a locus-specific probe can be generated, allowing allelism to be determined with certainty and permitting extension of studies to other pedigrees in order to confirm or exclude linkage.

Although lack of large extended pedigrees might limit the utility of DNA fingerprinting in the analysis of recessively inherited traits, it would not render it entirely useless. If consanguinous matings can be studied then the method of homozygosity mapping could be employed (27). The principle of this technique can be summarized thus: if two related individuals both carry a rare recessive gene, then (depending on the gene frequency, and how closely the individuals are related) they are likely to have inherited that gene from a common ancestor (28). Moreover, if the individuals are distantly enough related, then the proportion of their genomes which they share will be small. Furthermore, as the allelic frequencies of fingerprint bands (particularly of large bands) are very small, the probability of bands shared by the related individuals not having come from the common ancestor is low. If the shared fingerprint band is inherited in a double dose by children (of this consanguineous mating) who exhibit phenotypically the ASR trait, and in single or zero dose by unaffected children, then this provides support for the hypothesis of physical linkage between the trait locus and the band. The odds in favour of linkage can be calculated if allelism is assumed but, again, the band must be cloned if linkage is to be confirmed or excluded.

5.2 Statistical considerations

On a blot of average quality, the inheritance of $10-20$ fingerprint bands can be studied for each of the fingerprinting probes, a total of $30-60$ markers altogether. Genetic data suggest that there is no extensive clustering of HVR loci (16) and so it may be valid to assume that they are independently assorted. The probability of at least one of 45 independently assorted marker loci lying within a genetic distance of 15% of a given trait locus can be estimated by the algorithm of Elston and Lange (29,30) at P approximately 0.33. Given adequate pedigree, then, the odds are good for finding linkage using this method.

It is necessary to define a critical value for odds in favour of linkage so that when these odds are calculated for a preliminary fingerprint screening a decision can be made about going on to the next stage, of cloning the candidate band. In classical likelihood methods of analysis the critical value of the lod score is $z(\theta) = 3$; lods exceeding 3 are taken as evidence of linkage (31). Considering that the investment of effort required for the cloning of a fingerprint band is large, it seems sensible to require a stringent critical value to be exceeded on preliminary fingerprint screening before proceeding to the cloning step, for example odds of 1000:1 or better in favour of linkage. How can these odds be calculated?

It is simple to obtain a numerical estimate of the recombination fraction between the trait locus and a fingerprint marker.

$$\theta = \frac{r}{n}$$

where r = number of recombinants, n = total number of opportunities for recombination.

From this, an odds ratio can be calculated in the usual way, with

$$L(\theta) = \theta^r (1-\theta)^{n-r}$$

$$L(0.5) = 0.5^n$$

$$\text{and } \lambda = \frac{L(\theta)}{L(0.5)} = 2^n \theta^r (1-\theta)^{n-r}$$

This ratio does not account for the prior probability of obtaining a linkage result if multiple markers are examined. There is some controversy as to whether any correction should be made for multiple comparisons when fewer than 100 markers are tested. Ott (31) argues that while the chance of obtaining a false positive linkage result increases with the number of markers tested so also does the chance that one of the markers is truly linked to the trait locus under study, and, moreover, the latter probability increases more rapidly than the former, so that the proportion of true linkages among all linkage results does not decline. While this argument is mathematically convincing, it is counter-intuitive and therefore not entirely satisfactory. In linkage analysis a spuriously high initial odds ratio will decline into insignificance in the absence of linkage. A high rate of preliminary false positives would render this fingerprinting approach inefficient and must be avoided, but it is more critical to minimize the rate of false negatives and for this reason it may be reasonable to maintain this threshold for proceeding to the second stage of analysis, at least until a more convincing model is available for the problem.

5.3 Practical aspects of segregation analysis

5.3.1 *Minimum pedigree resources*

Analysis of the segregation of fingerprint bands is most straightforward in single sibships (*Figure 3*). However, in order for such an analysis to have a chance of yielding odds in favour of linkage which exceed the critical value of 1000:1, at least 10 meioses must be studied (when $n = 10$ and $r = 0$, $\theta = 0$ and $\lambda = 1024$, but when $n = 9$ and $r = 0$, $\lambda = 512$). Sibships of this size are not often available for study and frequently it is necessary to analyse segregation through an extended pedigree.

The frequency of band-sharing between unrelated individuals for fingerprint bands larger than 3 kbp is low. Thus, it is not vitally important for samples from both parents of a sibship to be available for study; given one parent's fingerprint, the other's can be reconstructed with a fair degree of certainty. If DNA from both parents is available, then these samples should be run on adjacent tracks on the gel. In this way the origin of bands in the offspring which happen to migrate similarly can be assigned most easily to a parent.

5.3.2 *Example of fingerprint segregation analysis*

An example is given here of analysis of the inheritance of fingerprint bands in a large single sibship (*Figure 3*).

Because of the inexactness of agarose gel electrophoresis under these conditions, which give rise to the 'banana effect', it is impractical to attempt digitization of fingerprint data. Instead, it is necessary to 'score' the fingerprints by hand. This is most easily done if one sibship is examined at a time, with the segregation of each parent's bands scored in turn. It is convenient to assign arbitrary letters to one parental phenotype, and numbers to the other.

Once the segregation of bands has been scored and tabulated for each sibship, the results can be analysed for evidence of linkage. If a hypothesis of linkage of the trait locus under study to a particular band in an affected ancestor is taken, then recombinant and non-recombinant individuals can be counted, a numerical estimate of the recombination fraction made, with $\theta = r/n$, and the odds ratio calculated as in Section 5.1. This procedure is carried out for each of the bands in the affected ancestor's fingerprint.

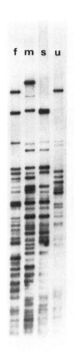

Figure 4. Determination of true paternity using DNA fingerprinting. Each band detected by minisatellite probe 33.15 in the DNA of individual **s** which cannot be scored in his mother, **m**, is present in his true father, **f**, while only three out of the eight non-maternal bands in **s** are present in a paternal uncle, **u**.

6. OTHER USES OF DNA FINGERPRINTING

6.1 Verification of pedigree structure

Frequently in genetic studies it is necessary to determine with certainty the true biological paternity of an individual. Traditional means by which this determination is made include the use of blood groups and protein and DNA markers. These suffer from the disadvantage of lacking a high degree of individual specificity, and therefore only negative associations can be made with certainty using these tests. By contrast, DNA fingerprints are virtually completely specific to the individual and so lend themselves well to the task of positive identification of an individual (17,32,33).

The bands comprising a DNA fingerprint are inherited stably in a Mendelian fashion. All of the bands in an individual's fingerprint should, then, be identifiable also in the fingerprint of either his father or his mother, allowing for a mutation rate of 10^{-4}/kb/meiosis. Approximately one half of the fingerprint bands in an offspring will be of paternal origin. If all three probes are used, an average of about 15 paternally-derived fragments in the $4-20$ kb range will be resolvable. The probability that the putative father will possess all 15 of these fragments by chance, and that incorrect paternity will not be detected, is exceedingly low ($<4 \times 10^{-11}$ if the putative father is unrelated to the true father, $<4 \times 10^{-5}$ if they are brothers) (*Figure 4*).

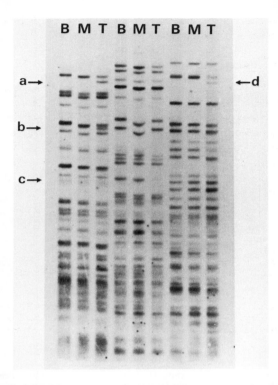

B M T B M T B M T

a→   ←d

b→

c→

Figure 5. The use of fingerprinting to compare DNA from neoplastic and normal tissues. The autoradiograph shows an analysis of DNA extracted from samples of peripheral blood (**B**), as well as normal mucosa (**M**) and tumour tissues (**T**) taken from three patients suffering from gastrointestinal malignancies. DNA (10 μg) was digested with *Hin*fI and run on a 1.0% agarose gel at 2 V/cm for 48 h then transferred to a nylon membrane and probed with minisatellite 33.15. In patient 1 two aberrant bands (**a,b**) are present in the tumour fingerprint. In patient 2 a band present in both peripheral leukocyte DNA and normal colonic mucosa DNA (**c**) is absent in DNA extracted from a colonic carcinoma. In patient 3 a novel band (**d**) has appeared in the fingerprint of DNA extracted from a gastric carcinoma. There are, as expected, no instances of difference between the normal leukocyte and normal mucosa fingerprints. In the three examples of the appearance of novel bands, a higher molecular weight band has lost intensity, suggesting that the new band has arisen in a subpopulation of the tumour cells by an unequal cross-over which has led to a loss of repeat copies from, and thus shortening of, the parent allele.

6.2 Comparisons within an individual

DNA fingerprinting provides a convenient means for rapidly screening the genome for somatic changes. This idea has found application in comparisons of leukocyte DNA to constitutional DNA following bone marrow transplantation and of DNA from tumorous and normal tissues (34,35).

In a high proportion of malignant tumours, DNA fingerprint changes, comprising appearance of new bands, or loss of bands, are found (*Figure 5*). This latter category of alterations is perhaps the more interesting, considering recent demonstrations of pathogenetically relevant losses of alleles in numerous cancers. If the probes described here are used to generate DNA fingerprints for normal and tumour DNA from an individual, then as many as 60 alleles in the size range 2−20 kb can be examined, giving a good chance of detecting alterations.

7. ACKNOWLEDGEMENTS

I would like to thank Martin Fey for allowing me to use the autoradiograph shown in *Figure 5* and for his critical reading of the manuscript, Rachel Kitt and Helen Blaber for their assistance in preparing the manuscript, Robin Sherrington for helpful discussion and Stephen Reeders for advice and encouragement. I thank the Rhodes Scholarship Trust for supporting me during this work and Professor Sir David Weatherall in whose department this work was carried out.

8. REFERENCES

1. Wyman,A.R. and White,R. (1980) *Proc. Natl. Acad. Sci. USA,* **77**, 6754.
2. Higgs,D.R., Goodbourn,S.E.Y., Wainscoat,J.S., Clegg,J.B. and Weatherall,D.J. (1981) *Nucleic Acids Res.,* **9**, 4213.
3. Bell,G.I., Selby,M.J. and Rutter,W. (1982) *Nature,* **295**, 31.
4. Capon,D.J., Chen,E.Y., Levinson,A.D., Seeburg,P.H. and Goeddel,D.V. (1983) *Nature,* **302**, 33.
5. Goodbourn,S.E.Y., Higgs,D.R., Clegg,J.B. and Weatherall,D.J. (1983) *Proc. Natl. Acad. Sci. USA,* **80**, 5022.
6. Proudfoot,N.J., Gill,A. and Maniatis,T. (1982) *Cell,* **31**, 553.
7. Jarman,A.P., Nicholls,R.D., Weatherall,D.J., Clegg,J.B. and Higgs,D.R. (1986) *EMBO J.,* **5**, 1857.
8. Higgs,D.R., Wainscoat,J.S., Flint,J., Hill,A.V.S., Thein,S.L., Nicholls,R.D., Teal,H., Ayyub,H., Peto,T.E.A., Falusi,Y., Jarman,A.P., Clegg,J.B. and Weatherall,D.J. (1986) *Proc. Natl. Acad. Sci. USA,* **83**, 5165.
9. Simmler,M.C., Johnsson,C., Petit,C., Rouyer,F., Vergnaud,G. and Weissenbach,J. (1987) *EMBO J.,* **6**, 963.
10. Smith,G.P. (1976) *Science,* **191**, 528.
11. Goodbourn,S.E.Y., Higgs,D.R., Clegg,J.B. and Weatherall,D.J. (1984) *J.. Mol. Biol. Med.,* **2**, 223.
12. Reeders,S.T., Breuning,M.H., Davies,K.E., Nicholls,R.D., Jarman,A.P., Higgs,D.R., Pearson,P.R. and Weatherall,D.J. (1985) *Nature,* **317**, 542.
13. Hodgkinson,S., Sherrington,R., Gurling,H., Marchbanks,R., Reeders,S., Mallet,J., McInnis,M., Petursson,H. and Brynjolfsson,J. (1987) *Nature,* **325**, 805.
14. Jeffreys,A.J., Wilson,V. and Thein,S.L. (1985) *Nature,* **314**, 67.
15. Nakamura,Y., Leppert,M., O'Connell,P., Wolff,R., Holm,T., Culver,M., Martin,C., Fujimoto,E., Hoff,M., Kumlin,E. and White,R. (1987) *Science,* **235**, 1616.
16. Jeffreys,A.J., Wilson,V., Thein,S.L., Weatherall,D.J. and Ponder,B.A.J. (1986) *Am. J. Hum. Genet.,* **39**, 11.
17. Jeffreys,A.J., Wilson,V. and Thein,S.L. (1985) *Nature,* **316**, 76.
18. Vassart,G., Georges,M., Monsieur,R., Brocas,H., Lequarre,A.S. and Christophe,D. (1987) *Science,* **235**, 683.
19. Sealy,P.G. and Southern,E.M. (1982) In Rickwood,D. and Hames,B.D. (eds), *Gel Electrophoresisis of Nucleic Acids—A Practical Approach.* IRL Press, Oxford, p. 39.
20. Old,J.M. and Higgs,D.R. (1982) In *Methods in Haematology.* Weatherall,D.J. (ed.), Churchill Livingstone, London, p. 74.
21. Singh,L. and Jones,K.W. (1984) *Nucleic Acids Res.,* **12**, 5627.
22. Davies,K.E., Pearson,P.L., Harper,P.S., Murray,J.M., O'Brien,T., Sarfarazi,M. and Williamson,R. (1983) *Nucleic Acids Res.,* **11**, 2303.
23. Monaco,A.P., Bertelson,C.J., Middlesworth,W., Colletti,C.-A., Aldridge,J., Fischbeck,K.H., Bartlett,R., Pericak-Vance,M.A., Roses,A.D. and Kunkel,L.M. (1985) *Nature,* **316**, 842.
24. Cavenee,W.K. (1986) *Trends Genet.,* **2**, 299.
25. Botstein,D., White,R.L., Scolnick,M.H. and Davis,R.W. (1980) *Am. J. Hum. Genet.,* **32**, 314.
26. Wong,Z., Wilson,V., Jeffreys,A.J. and Thein,S.L. (1986) *Nucleic Acids Res.,* **14**, 4605.
27. Lander,E.S. and Botstein,D. (1987) *Science,* **236**, 1567.
28. Bulmer,M.J. (1985) *The Mathematical Theory of Quantitative Genetics.* Oxford University Press, Oxford.
29. Elston,R.C. and Lange,K. (1975) *Ann. Hum. Genet.,* **38**, 341.

30. Elston,R.C. (1975) *Hum. Gene Mapping,* **2**, 290.
31. Ott,J. (1985) *Analysis of Human Genetic Linkage.* Johns Hopkins University Press, Baltimore, Maryland.
32. Jeffreys,A.J., Brookfield,J.F.Y. and Semeonoff,R. (1985) *Nature,* **317**, 818.
33. Gill,P., Jeffreys,A.J. and Werrett,J.D. (1985) *Nature,* **318**, 577.
34. Thein,S.L., Jeffreys,A.J. and Blacklock,H.A. (1986) *Lancet,* **ii**, 37.
35. Thein,S.L., Jeffreys,A.J., Gooi,H.C., Cotter,F., Flint,J., O'Connor,N.T.J., Weatherall,D.J. and Wainscoat,J.S. (1987) *Br. J. Cancer,* **55**, 353.

CHAPTER 8

Mapping complex genetic traits in humans

ERIC S.LANDER

1. INTRODUCTION

Human genetics has come of age in the past few years. Of course, it has been clear since the re-discovery of Mendel that humans obey the same laws of heredity as other organisms. Indeed, studies of the inheritance of genetic diseases in human families provided some of the first persuasive evidence for Mendelism. Despite this promising start, however, human genetics has lagged ever since far behind the study of organisms such as *Drosophila* and maize for two fundamental reasons:

(i)   because *Homo sapiens* is not a laboratory animal in which matings could be arranged to suit experimental purposes; and

(ii)   because few genetic markers had been found which were heterozygous frequently enough to allow existing mating in natural populations to be informative for linkage analysis.

With the advent of recombinant DNA technology came the suggestion (1) that poly-morphisms in DNA sequence—which could be conveniently visualized as restriction fragment length polymorphisms or RFLPs—would be common enough that they could be used as generally informative genetic markers allowing the systematic study of human heredity, including the construction of a true linkage map of the human genome.

The application of RFLP technology to simple Mendelian diseases has proceeded rapidly. Extensive studies have confirmed that RFLPs are indeed abundant (2) and, more recently, methods have been devised for finding highly polymorphic RFLPs (3). To date, over 2500 RFLPs have been identified (4). Screening of randomly chosen RFLPs in affected pedigrees has identified DNA markers showing close genetic linkage to Duchenne muscular dystrophy (5), Huntington's disease (6), cystic fibrosis (7 − 10), adult polycystic kidney disease (11), retinoblastoma (12), early-onset familial Alzheimer's disease (13), bipolar affective disorder among a large Amish kindred (14), von Recklinghausen neurofibromatosis (15,16), multiple endocrine neoplasia type 2a (17,18) and familial polyposis (19). Moreover, the rapid progress in the construction of a human genetic linkage map (20,21) promises to make the entire process more efficient, by replacing the use of randomly-chosen RFLPs with the systemic screening of RFLPs spaced throughout the genome.

These successes make clear that RFLPs can be found linked to any human disease, provided that it:

(i)   shows simple Mendelian recessive or dominant transmission;

(ii)   is caused by mutations at a single genetic locus; and

(iii)   is common enough to allow the collection of a sufficient number of families with multiple affected members.

Unfortunately, much of what we want to know about human heredity concerns traits whose underlying genetics are less favourable for analysis. The problems include:

(i)   *incomplete penetrance*, in which some individuals carrying the mutant genotype fail to express the mutant phenotype;

(ii)   *phenocopies*, in which individuals of normal genotype may display the mutant phenotype due to non-genetic causes;

(iii)   *genetic heterogeneity*, in which mutations at several different genetic loci may result in apparently identical clinical conditions;

(iv)   *gene interactions*, in which phenotypes result from the interaction of alleles at more than one locus; and

(v)   *rarity*, which may make it difficult or impossible to collect many families with multiple affected individuals.

These problems are common in genetically well-studied organisms, such as bacteria, yeast, nematodes and fruit flies. Evidence is accumulating that humans are no different. Geneticists usually surmount these problems by isolating pure-breeding strains carrying a mutation at a single locus, propagating large numbers of individuals under controlled conditions and arranging crosses at will. Human geneticists, however, must take crosses as they find them.

In this chapter, we give an overview of the general considerations involved in planning a human linkage study when such complications may arise. The emphasis will be on the practical issues of choosing families in such a way as to maximize the chance of detecting linkage. We shall not discuss detailed calculation of required sample sizes, nor the mathematical aspects of linkage analysis. [For more information on these topics, see (22–24) upon which this survey draws heavily.] The optimal design of a linkage study, it should be noted, depends on the precise mode of inheritance of the disease (degree of penetrance; number of independent loci; nature of genetic interactions; frequency of phenocopies; frequency of families with multiple affected individuals etc.), which is rarely known in advance. If the genetics are in fact too complex—say, if the disease could be caused by mutations at any one of 100 loci—linkage studies will obviously not succeed. An investigator's choice of design must therefore depend, to some extent, on informed guesses about the underlying complexities. Despite this caveat, careful planning will substantially increase the power of linkage studies to analyse diseases of complex inheritance.

2. BASICS OF LINKAGE ANALYSIS

A comprehensive introduction to the fundamentals of genetic linkage analysis has been given in a chapter in this series by Ott (25). We shall simply summarize the basic notion of likelihood analysis here.

Suppose that we wish to compare a 'test' hypothesis, H, to an alternative 'null' hypothesis, H_0. For example, H might be the hypothesis that a RFLP locus is linked to a disease at 10% recombination, while H_0 might be the hypothesis that they are unlinked. (Actually, this is a slight oversimplification: we should consider many alternative hypotheses H, corresponding to linkage at various different recombination fractions.) After collecting data, we would compute two probabilities: P, the chance that the observed data would have arisen if the hypothesis H is true, and P_0, the chance that it would have arisen if the null hypothesis H_0 is true. The odds ratio, P/P_0, measures how much more likely the data are to have arisen under H than H_0. If the odds ratio exceeds some sufficiently large threshold T, we reject H_0 in favour of H.

For convenience, human geneticists prefer to report the $\log_{10}$ of the odds ratio, called the *LOD score*. The odds ratios from independent samples may be combined by multiplying them together, and thus the LOD scores from independent samples may simply be added.

In designing an experiment, there are two relevant questions:

(i)   Threshold—how large should the threshold T be in order to make the risk of incorrectly rejecting H_0 (when it is in fact true) sufficiently small?

(ii)   Required sample size—how large a sample will be needed in order to make the chance of accepting H (when it is in fact true) sufficiently high?

In human genetics, the accepted threshold (26) for simple detection of linkage is an odds ratio of at least 1000:1, corresponding to a LOD score of at least 3.0. This threshold is less stringent than it might appear. Given two randomly chosen loci in the human genome, they are very likely to be on different chromosome arms. Roughly, the odds are 50:1 against linkage. Even though the data may be 1000 times more likely to have arisen under linkage than under non-linkage, the fact that non-linkages are 50 times more common than linkages implies that a LOD score of 3 corresponds to only about 20:1 odds in favour of linkage. That is, a LOD score of 3 will prove spurious about one time in 20. (Occasionally, an author will erroneously claim that a LOD score of 3 means that linkage is 1000 times more likely than non-linkage; it does not.) A LOD score of 4 will be spurious about one time in 200.

The required sample size can be estimated by calculating the expected contribution that each family will make to the total LOD score. Given the structure of the family, one computes the expected LOD score, or ELOD, that will be obtained. One should collect enough families so that the sum of the ELODs substantially exceeds the threshold needed for declaring linkage.

The strength of the likelihood method is its flexibility. One can study more complex genetic transmission simply by formulating more complex hypotheses. To deal with genetic interactions, for example, we could examine the hypothesis that mutations at two different loci are *both* required in order to produce a disease. Or, to deal with genetic heterogeneity, we could examine the hypothesis that mutations at either one of two loci could suffice to cause the disease, in which case each family would show linkage to one (but not both) of the loci.

3. THE HUMAN GENETIC LINKAGE MAP

Human genetics will become substantially more powerful as a complete linkage map of the human genome becomes available. Such a map will consist of RFLP markers spaced evenly along the chromosomes, a total of about 3300 centiMorgans (cM).

Great progress has been made toward such a linkage map in the past few years. In late 1987, Donis-Keller *et al.* (20) reported the construction of a 403-locus map of the human genome with linkage groups on all chromosomes and White *et al.* (21) distributed a booklet containing a 255-locus map with linkage groups covering 17 of 23 chromosomes. Since the overlap between the markers studied by the two groups is small, a map of about 600 loci emerges when these maps are pooled. Since many more RFLPs are currently being mapped by these and other groups, it seems quite realistic to suppose that a map of over 1000 RFLPs will soon be available. On average, such a map will have markers roughly every 3 cM (although the spacing will be random, not regular).

Such a dense RFLP map of the human genome offers a number of advantages.

(i) *All families will become fully informative for linkage analysis.* When random, unmapped RFLPs are used in linkage analysis, meioses occurring in individuals homozygous for the RFLP contribute no information at all to the two-point linkage analysis of the RFLP and the disease under study. The availability of a complete map, however, makes it possible to choose several nearby RFLPs in a region so as to ensure that virtually all individuals are heterozygous for at least one of the RFLPs. Extracting this full information requires multi-point linkage analysis, which can be performed by computer.

(ii) *Inheritance at a disease locus can be followed more accurately through the use of flanking markers.* If no recombination is observed between markers on either side of a putative disease locus, then one can be sure that the disease locus between them has been co-inherited—except for the small chance that a double cross-over has occurred. Indeed, geneticists have long recognized that such three-point crosses are inherently more precise than two-point crosses. The extra precision makes it easier to prove or disprove hypotheses, and thus fewer families are needed (see, for example *Figure 1*).

(iii) *Inheritance of several regions of the genome can be followed simultaneously.* As we note below, this makes it possible to detect linkage to heterogeneous genetic diseases.

In the future, we envisage that linkage studies will initially employ a standard battery of perhaps 150−200 highly informative RFLPs distributed throughout the genome, about one every 15−20 cM. Once the pedigrees under study have been genotyped for each of these loci, a computer analysis will be performed to test each genomic interval for suggestive evidence of linkage (e.g. a LOD score of 1.5−2.0). Any interval showing suggestive evidence for linkage will then be studied with a higher density of RFLPs, perhaps one every 3 cM. This will extract the full information available from the pedigrees, and will increase the LOD score. Accordingly, the relevant question in planning a linakge study will be: how many families are needed to be able to detect a hint of linkage given a sparse RFLP map and to be able definitively to prove linkage given a dense RFLP map?

4. PLANNING A LINKAGE STUDY OF A DISEASE WITH SIMPLE INHERITANCE

By simple inheritance, we mean that a disease is (i) inherited as a classical Mendelian dominant or recessive trait showing complete penetrance and (ii) is caused by mutations at a single locus. A careful study of the epidemiology of a disease will usually reveal whether condition (i) holds. By contrast, condition (ii) usually must be accepted on

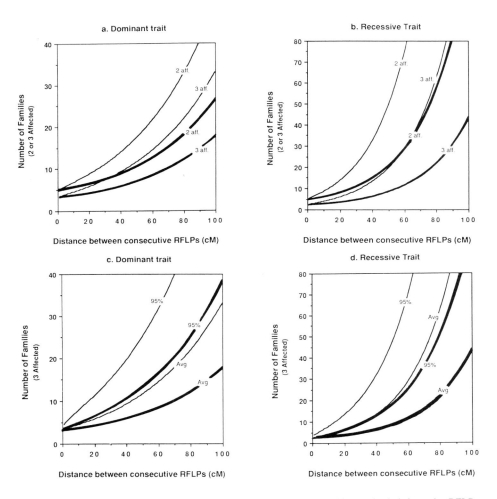

Figure 1. The number of families needed to map simple Mendelian traits with completely informative RFLPs—comparison of single-marker method (thick curves) with interval mapping (thick curves). Phase is assumed to be known for the RFLP markers and for a dominant trait, but the phase of a recessive trait is assumed to be unknown. For a dominant trait, the graphs show the expected number of meioses needed to reach a LOD score of 3.0 (**a**) and the number needed to ensure a 95% certainty of reaching a LOD of 3.0 (**c**). For a recessive trait, the graphs show the expected number of families of different types (classified by the number of affected sibs) to reach a LOD score of 3.0 (**b**) and the number needed to ensure a 95% certainty (**d**). The number of affected individuals in the family is relevant because the phase of the trait is presumed to be unknown in the recessive case; sibs thus contribute information about phase. In the phase-known dominant case, meioses can be treated independently.

faith: no powerful method (short of actually mapping the disease gene) is available for *proving* that mutations at only a single locus are involved—although there are ways to cast doubt on the proposition. We return to this issue below, in our discussion of genetic heterogeneity (Section 6).

Huntington's disease and cystic fibrosis show clear Mendelian dominant and recessive inheritance, respectively. As it turns out, all cases result from mutations at a single locus (on chromosome 4p for Huntington's and on 7q for CF), although this became clear only once linkage was established.

A disease with simple inheritance is relatively easy to linkage map. By computing the ELOD (expected LOD score), one can straightforwardly estimate the required sample size, *Figure 1a,b* shows the number of nuclear families with either two or three affected required to assure a 50% chance of success in mapping a dominant or recessive trait, as a function of the density of RFLPs used. The figure shows the number required (i) if the RFLPs are used one at a time and (ii) if the power of a RFLP map is exploited through the use of flanking markers. *Figure 1c,d* shows the number of families required to assure a 95% chance of success. (In each case the RFLPs are assumed to be completely informative; roughly twice as many families would be needed if one used RFLPs which were informative only half the time.) As *Figure 1* makes clear, the number of families required is quite manageable (<12) given a map of RFLPs spaced every $20-30$ cM throughout the human genome.

We next consider traits showing more complex inheritance and we discuss the problems that arise in linkage mapping and ways to overcome them.

5. COMPLEXITIES: INCOMPLETE PENETRANCE AND MISDIAGNOSIS

A genetic trait that requires additional factors to become manifest will show incomplete penetrance. These factors might include environment, genetic background or chance. A simple example is retinoblastoma, in which individuals inheriting one damaged copy of a recessive oncogene develop cancer if a somatic event inactivates the second copy; this chance event is quite frequent, so penetrance is high (12). In families with inherited forms of Alzheimer's disease, the disease may not show complete penetrance simply because some members die of other causes before they display any symptoms. Wernicke–Korsakoff syndrome seems to be due to a mutation causing transketolase to bind thiamine pyrophosphate less avidly than normal, but a clinical phenotype (alcohol-induced encephalopathy) is apparent only in patients with a dietary thiamine deficiency (27); thus, the degree of penetrance can vary widely with environment and with the precise definition of the phenotype (encephalopathy versus thiamine pyrophosphate binding). Many diseases, such as inherited forms of manic depression, show incomplete penetrance for reasons that are not understood.

In a dominant disease with complete penetrance, unaffected children of an affected parent are as informative as affected children: in both cases, we can be certain which allele at the disease locus must have been inherited. When penetrance is incomplete, however, the genotype of an unaffected descendant of affected individuals is always in doubt. Not surprisingly, relatively little can be learned by studying the markers that they have inherited, since we cannot even be sure whether they inherited the disease allele. *Figure 2* shows how quickly the contribution to the LOD score plummets with incomplete penetrance.

The problem of uncertain genotypes for an unaffected individual is evident even for a recessive trait of *complete* penetrance: there are three possible genotypes for an unaffected sibling, but only one for an affected sibling. For a tightly linked marker, the odds ratio will be 1:(3/4) for an unaffected sibling and 1:(1/4) for an affected sibling, making the contribution to the LOD score only *one-fifth* as large for unaffected as for

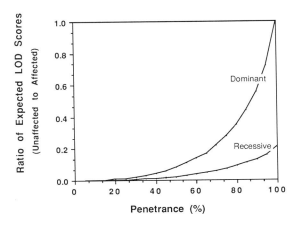

Figure 2. Relative values for linkage mapping for affected and unaffected sibs as a function of penetrance of disease gene. Specifically, the graph shows the ratio of the expected LOD (ELOD) for an unaffected sib to that for an affected sib, when a marker is available at 0 cM. All phases are assumed to be known.

affected siblings. The situation is only exacerbated by incomplete penetrance (*Figure 2*).

In reckoning the value of a family for linkage analysis, one should focus on the number of meioses which give rise to affected individuals (except in the case of a dominant disease of complete penetrance, in which case all meioses in which the disease locus is heterozygous are valuable). There is no harm in collecting unaffected siblings, but one should be clear about the fact that unaffected siblings contribute relatively little to the search for the disease locus—except in the case of a dominant disease of nearly complete penetrance. (We should note that unaffected siblings, although not very informative about the disease locus, may be valuable for inferring the genotype of a dead parent at the RFLP marker loci.)

Incomplete penetrance, it should be noted, poses one grave risk for linkage analysis. If an adequate degree of incomplete penetrance is not built into the models to be tested in the linkage analysis, then true linkage might be missed entirely: unaffected individuals will appear to be recombinants. In order rigorously to exclude linkage of a disease with incomplete penetrance to a particular region, one should perform a linkage analysis in which the degree of penetrance is assumed to be very low. The resulting LOD score will reflect the information deriving from affected individuals. The LOD score will decrease, making apparent the extent to which the conclusion relies on apparent recombinants among the unaffecteds.

Reciprocal to the problem of incomplete penetrance is the issue of potential misdiagnosis of unaffected individuals as affected. Whereas incomplete penetrance renders the genotype of unaffected individuals ambiguous and thereby decreases their value to linkage analysis, the possibility of misdiagnosis means that the genotype of *affected* individuals must be considered uncertain as well. Misdiagnosis may be unlikely for clear disorders such as cystic fibrosis, but it is a serious matter for psychiatric disorders: a wide range of behaviours are included among the diagnostic criteria for such disorders, and the existence of affected close relatives is often taken as confirmatory evidence! Using *Figure 3*, one can estimate the effect of a given rate of misdiagnosis

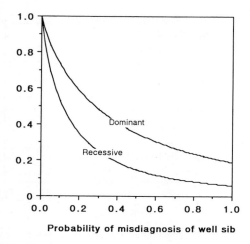

Figure 3. The curves show the relative ELOD for a phase-known meiosis producing an affected child, in the case of dominant and recessive diseases, given the probability that a genetically normal sibling would be misdiagnosed as being affected.

on a linkage analysis. Because misdiagnosis can destroy the prospects for linkage analysis, the best policy is to designate any individual for whom the diagnosis is in doubt as 'phenotype unknown'. After linkage has been found, one can return to such individuals and discover whether they have inherited the disease genotype. In this way, one can discover whether the ambiguous symptoms are in fact a manifestation of the same disease entity.

6. COMPLEXITIES: GENETIC HETEROGENEITY AND PHENOCOPIES

The most serious obstacle to linkage studies is genetic heterogeneity, the situation of a phenotype that can be caused by mutations at any one of several loci. The classic paradigms of genetic heterogeneity, well-known in lower organisms, are (i) the interruption of a single biochemical pathway at any of its steps and (ii) the loss or disruption of a heteromultimeric protein complex by mutation in the structural gene for any subunit, although other possibilities are not hard to imagine (such as regulatory genes or post-translational modifications). In lower organisms, genetic heterogeneity of phenotypes is the rule, although it causes little confusion because geneticists are able to arrange simple complementation crosses to test for allelism between mutations.

Although the situation has been less studied in humans, there is no reason to think that genetic heterogeneity is less common. For example, hereditary methaemoglobin-aemia, once thought to be a homogeneous clinical entity, can be produced by mutations in either the alpha or beta chains of haemoglobin or in NADH dehydrogenase (28). Elliptocytosis (29) and Charcot−Marie−Tooth disease (30,31) have been found to be genetically heterogeneous, because in each case tight linkage has been seen in some large pedigrees but is entirely absent in others. Similarly, inheritance studies in a large Amish pedigree (14) have demonstrated linkage between a region of chromosome 11p and an inherited form of bipolar affective disorder (i.e. manic depresssion), but this locus is not linked in several other large pedigrees (32,33). Xeroderma pigmentosa and ataxia telangiectasia are probably genetically heterogeneous, since *in vitro* assays on cell fusions reveal nine and five complementation groups, respectively (34−36). Natural complementation crosses can occasionally be studied, as when two albinos married and produced normal children (37), showing that albinism is genetically heterogeneous—as had been suspected from phenotypic distinctions and from population genetics (a higher rate of consanguinity among parents of affecteds than expected for a single gene). Similar evidence suggests heterogeneity for congenital deafness (28).

Heterogeneity is the geneticist's nightmare because evidence for linkage to a locus in one family will be offset by evidence against linkage in another family. Even a modest degree of heterogeneity may cause the traditional LOD score to be negative—even in the neighborhood of a marker tightly linked to one of the disease-causing loci. (Accordingly, one must be very cautious about the practice of 'excluding' a locus from a region whenever heterogeneity is possible, which is almost always.)

The potential solutions for overcoming heterogeneity are as follows:

(i) *Splitting the clinical phenotype.* The best way to overcome genetic heterogeneity is to discover a clinical phenotype which accurately distinguishes between the different genetic forms: the study of one heterogeneous disorder is then transformed into the study of two or more homogeneous disorders. For example, clinical phenotypes were used to distinguish two forms of neurofibromatosis (von Recklinghausen NF and Bilateral Acoustic NF), which recent mapping studies have proven map to different loci (15,16,38).

Unfortunately, there is no general recipe for correctly splitting a genetic disease. The clinical presentation of a disease may vary greatly for at least three reasons.

(a) A single mutation at a single locus may be variably expressed due to unlinked genetic modifier loci, due to environmental effects or due to chance.

(b) Different mutations at a single locus may disrupt the same biochemical function in varying ways or to a varying extent, as in Becker muscular dystrophy which is caused by less severe alleles of the Duchenne muscular dystrophy gene.

(c) Mutations at different loci may be involved.

These situations are referred to, respectively, as (a) background effects, (b) allelic heterogeneity and (c) genetic heterogeneity.

A disease may thus be clinically heterogeneous, even though it is not genetically heterogeneous, and vice versa. In order to regard a clinical distinction as a potentially useful tool for genetics, one must first show that the members of any given family all fall into the same phenotypic subclass. (Of course, there is no genetic way to tell— short of a complementation cross—whether such a distinction could result from either allelic or genetic heterogeneity.) For example, a natural candidate for genetic splitting of a psychiatric disorder might be responsiveness to a particular drug. Sadly, few studies have explored whether affected individuals in a family tend to show similar patterns of drug responsiveness.

Splitting phenotypes is an important quest of the medical geneticist. One must rely on inspiration or luck, however, and one cannot tell when the task is finished.

(ii) *Studying a single, large pedigree.* Another approach to avoiding genetic heterogeneity is to find a single pedigree large enough to suffice for linkage analysis. Gusella *et al.* (6) for example, analysed a single extended Venezuelan pedigree segregating Huntington's disease, containing hundreds of individuals. Although later linkage studies indicate that Huntington's is genetically homogeneous, this could not have been known in advance. By studying a large Amish pedigree segregating for an inherited form of bipolar affective disorder Type I (severe manic depression), Egeland *et al.* (14) discovered a linkage to RFLP markers on chromosome 11p. Several other large pedigrees (32,33) do not show linkage to this region, demonstrating that heterogeneity is indeed present.

There are several drawbacks to the large pedigree approach:

(a)   For many dominant diseases and almost all recessive diseases, sufficiently large pedigrees will simply be impossible to find.

(b)   The attempt to 'expand' a pedigree by identifying as many affected members as possible may actually *introduce* heterogeneity in the case of relatively common complex diseases, such as manic depression, heart disease, etc. In assembling such pedigrees in a large mixed population such as the general USA population, it becomes likely that portions of the pedigree may segregate for different diseases alleles or loci. From the standpoint of linkage analysis, such intra-pedigree heterogeneity is much worse than inter-pedigree heterogeneity.

(c)   Even when the large pedigree approach succeeds, the conclusion may not apply to the general population.

Further extensive studies in many other (necessarily smaller) families will be needed in order to know whether the finding is general. Indeed, the linkage between manic depression and markers on chromosome 11p found in the large Amish pedigree mentioned above has yet to be found in any other pedigrees. Conceivably, it could be quite a rare form of manic depression. Despite these limitations, the large pedigree approach can be quite valuable, especially in the case or diseases which are not very common.

(iii) *Studying a single geographic region.* A related approach to avoiding heterogeneity is to restrict attention to a single population which has been relatively isolated from a genetic standpoint. Diseases which are heterogeneous in a mixed population such as the USA are more likely to be caused by a single mutation in such an isolated population.

(iv) *Simultaneous search of the genome, using smaller families.* The availability of a complete RFLP linkage map of the human genome makes it possible to detect linkage even when more than one locus is segregating in the population under study. This recently developed method (23), known as *simultaneous search*, amounts to a natural generalization of the standard approach to linkage mapping.

After preparing DNAs from a collection of families, one would screen RFLPs spanning the entire genome. If a single genetic interval shows tight linkage with the disease in all families, the mapping would be finished. If no single genetic interval displayed tight linkage, one would then examine all *pairs* of intervals, searching for a pair of intervals A and B with the following property: every family either showed tight linkage to interval A or else tight linkage to interval B. If the disease is caused by mutations at either of two loci, then the correct two intervals will clearly display the desired property.

a. How often does each interval segregate away from the disease?

Interval	A	B	C	D	E	F	G	H	I
Crossovers "Observed"	11	6	8	13	10	4	12	8	9
Crossovers Expected:									
if trait in B	10	0	10	10	10	10	10	10	10
if trait in F	10	10	10	10	10	0	10	10	10
if in B or F	10	5	10	10	10	5	10	10	10

Total Crossovers Possible: 20

b. How often do both B and F segregate away from the disease?

	A	B	C	D	E	F	G	H	I
A		3	5	7	5	3	9	5	5
B			3	4	5	0	5	3	3
C				3	4	2	5	3	4
D					7	3	12	7	8
E						2	8	4	5
F							3	2	2
G								7	12
H									4

c. Number of families needed to map a heterogeneous dominant trait

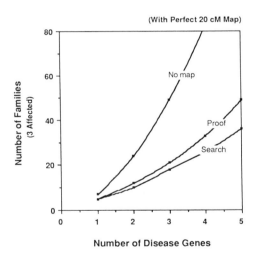

(With Perfect 20 cM Map)

No map

Proof

Search

Number of Families (3 Affected)

Number of Disease Genes

Figure 4. Illustration of the principle behind the simultaneous search method and the power of this approach for mapping heterogeneous traits. In (**a**) a hypothetical genome is divided into nine intervals (A, B . . . I) flanked by completely informative RFLP markers. Twenty informative meioses are scored, resulting in the 'observed' data for crossovers with a dominant disease gene shown in the second line. The number of crossovers expected is shown, under three different hypotheses about the location of the disease gene. The third hypothesis, that the disease is caused equally often by a gene in B and a gene in F, fits the data best, but the fit is not convincing. Instead, in (**b**) we record the number of meioses in which the trait is seen to recombine with *both* of a pair of intervals. Here, the pair of intervals B and F stand out more markedly. The number of families with three affected sibs required to map the gene causing a heterogeneous recessive trait or a heterogeneous dominant trait is shown in (**c**) and (**d**), respectively. In each case, the genes causing the trait are assumed to be equally frequent. The curve marked 'Search' pertains to proving that the set of loci shows linkage according to simultaneous search; the curve marked 'Proof' pertains to subsequently proving the involvement of each of the loci in the set. The curve marked 'No map' pertains to using admixture methods alone. Phase is assumed to be known for RFLP markers and for dominant traits but not for recessive traits.

To calculate the appropriate generalization of the LOD score for any pair of intervals A and B, one would compute (a) the chance that the family data would have arisen if the disease is caused by mutations in interval A and by mutations in interval B (with roughly equal frequency) and (b) the chance that the data would have arisen if these two intervals are unlinked to the cause(s) of the disease. (The method can be straightforwardly extended to allow three or more intervals, to allow a fraction of families with no genetic basis, or to allow different allele frequencies for different loci—in each case with some loss of power.) The key point is relatively simple: one can more easily recognize a pair of loci which account for segregation in all families, than one can a single locus that accounts for segregation in only half the families.

The only issue is the appropriate threshold for declaring linkage. Because many more hypotheses are being considered than in looking for linkage to a single interval, the *a priori* chance that any given hypothesis is correct is smaller and, thus, the usual threshold of 3.0 must be increased when two intervals are considered simultaneously. Lander and Botstein considered this mathematical issue in (23) and calculated the number of families which one would expect to need in order to detect linkage to a dominant or recessive heterogeneous trait, given families of different sizes (as an example see *Figure 4*). In brief, one should collect families with at least three (and preferably four) affected individuals in a single generation and, in the case of a dominant disease, an affected parent.

The method of simultaneous search can be used not only to find loci in the first instance, but to confirm a set of candidate loci found by other methods (the only difference being that a lower threshold is needed for confirmation than for initial detection). For some heterogeneous diseases, linkages to different loci may first be detected by traditional linkage analysis in large pedigrees. In order to determine whether these loci suffice to explain the disease in the general population, it will be necessary to use simultaneous search to study small families (no one large enough) to yield a LOD score of 3.0).

We should note that although genetic heterogeneity can wreak havoc with traditional linkage analysis, it is not even detectable by segregation analysis (the study of pedigrees to see if they fit Mendelian patterns of transmission). No matter how many independent loci can cause a recessive trait, each family segregates the expected Mendelian 3:1 ratio.

Closely related to the problem of genetic heterogeneity is the issue of phenocopies. A phenocopy is said to occur when non-genetic causes (such as a virus or environmental conditions) result in a phenotype which is also associated with an inherited condition. For example, homozygosity for a mutant allele of the α_1-anti-trypsin gene often causes emphysema, but the more common cause of this phenotype in the general population is cigarette smoking (28).

The existence of phenocopies has a similar effect to heterogeneity on linkage analysis: while some families will show linkage to a marker locus, others will segregate independently. Provided that phenocopies occur independently in the general population, one may be able to ensure that the families studied are 'genetic' by requiring a large number of affected individuals. Thus, a genetic aetiology is more likely in a family with five close relatives with breast cancer, then a family with two. However, this argument fails if phenocopies occur in familial clusters—for example, due to common environment. In this case, one must attempt to construct a homogeneous sample consisting of families segregating for a genetic form of the disease, by means of the strategies described above for genetic heterogeneity. When both genetic heterogeneity and phenocopies are possible, simultaneous search can be modified to take account of a fraction of 'unlinked' families.

In any case, too large a fraction of phenocopies in the sample will obscure any attempt to find linkage.

7. COMPLEXITIES: GENETIC INTERACTIONS

Some traits may result from genetic interactions between alleles at several loci. Examples abound in lower organisms, but have been hard to elucidate in humans. It is known, for example, that α-thalassaemia partly suppresses β-thalassaemia. Also, it is suspected that at least some of the diseases that show partial association with HLA genotype may involve other loci as well.

The most fundamental genetic interaction is a *synthetic trait*. Such a trait involves several 'component' loci. The trait results only when the appropriate mutant alleles are present at all of the component loci; a completely normal phenotype results otherwise. At some of the component loci, the alleles might act dominantly; at others, they may act recessively. Some may be rare, while others may occur at a high frequency in the population. The incidence of the trait will reflect the product of the frequencies of the mutant genotypes at the component loci.

One potential problem in mapping the loci involved in a synthetic interaction is that some parents may actually be homozygous for one of the component loci (although they will not express the trait because they lack the appropriate genotype at other component loci). Since an affected child can inherit either chromosome from such a parent, the trait will not show linkage to the locus in such a family. Analytically the effect is the same as if there were a large fraction of families having a non-genetic cause.

The problem of parental homozygosity will not be significant, however, unless the allele frequency at the component locus is rather high. In essence, one will only succeed in mapping those component loci at which the frequency of trait-causing alleles is low ($<10\%$).

It may be possible to take some measures to minimize the problem of parental homozygosity. For example, contrary to usual practice, it may be undesirable to study a family with too *high* a proportion of affected sibs: in such a family, it is more likely that one or both parents may be homozygotes for one or more of the component loci [(22), and E.S.Lander and D.Botstein, in preparation].

Leaving aside the problem of parental homozygosity, there is a further pitfall that must be avoided in mapping a genetic interaction: unaffected children will greatly confuse any attempt to map any single component locus. Even though a child inherits the disease-causing alleles at one component locus, he may not inherit the disease because he fails to inherit the required alleles at other loci. In essence, any one component locus causes the disease only with low penetrance. Thus, as noted above, the genotypes of unaffected children cannot be used as evidence against linkage to a candidate locus; one may only rely on the genotypes from affecteds.

Provided that one relies only on meioses producing affected individuals and that the

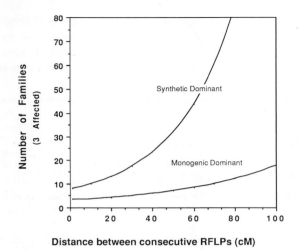

Figure 5. The number of families needed to map a dominant component allele involved in causing a synthetic trait via gene interaction. Which parent contributes the allele, as well as the phase of the allele, is assumed to be unknown. The phase of RFLP markers is assumed to be known. The allele is assumed to be present at <5% in the general population, so that one parent may be assumed to be a heterozygote and one parent homozygous normal (see text). For comparison, the number of families needed when a trait is caused monogenically by a dominant allele is also shown.

allele frequency is not too high, linkage analysis becomes relatively straightforward.

(i)   For a recessive component locus, the mapping proceeds just as in the monogenic case; the trait will co-segregate with each of the recessive component loci; and, the same number of families with a given number of affecteds is required.

(ii)   For a dominant component locus, the only difference from the monogenic case is that one does not know which of the two parents contributes the disease-causing allele. Because of this lack of information, one requires a larger number of families than would be needed in the monogenic case (*Figure 5*).

8. POLYGENIC INHERITANCE

Some traits are suspected to involve inheritance even more complex than genetic heterogeneity, incomplete penetrance or synthetic interactions. An extreme would be *polygenic inheritance*, in which alleles at a number of different loci interact in an additive fashion to determine one's risk of a trait or disease. Each additional 'bad' allele would increase one's predisposition, but no single locus would be essential to the aetiology of the disease. Thus, it is virtually impossible to infer genotype from phenotype. Calculations show that it would be impractical to map polygenes in the general human population.

Methods have been developed (E.S.Lander and D.Botstein, in preparation), however, for exploiting the full power of a complete RFLP linkage map to find polygenic factors in crosses between animal strains differing for a physiological trait. Such investigations might uncover specific loci that could then be studied more directly in humans.

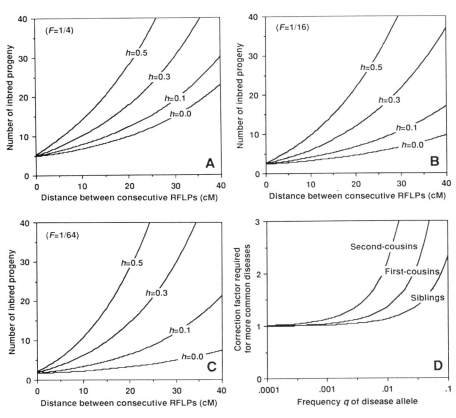

Figure 6. Number of inbred progeny needed to map a rare ($q = 0$) recessive disease via homozygosity mapping, as a function of the spacing between consecutive RFLPs and the degree of polymorphism of each RFLP. The four curves refer to RFLPs that are found homozygous with probability 50%, 30%, 10%, and 0% (limiting case) in the general population. (**A**) Progeny of a sibling mating; (**B**) progeny of a first-cousin mating; (**C**) progeny of a second-cousin mating. If the disease is not rare, the required numbers might be multiplied by the approximate correction factor in (**D**). (The correction factor given is exact when the distance between consecutive markers (d) $= 0$ and the rate of homozygosity (b) $= 0$; it differs, but only slightly, for other cases.)

9. MAPPING RARE RECESSIVE TRAITS

For many genetic diseases, it will be difficult to collect enough families with multiple affected individuals for traditional linkage analysis. Many recessive disorders of medical or biological importance are rare, occurring sporadically in isolated cases or small clusters. Even for a common disease, it may be difficult to collect families with multiple affecteds when the disease results in death at a very young age.

One potentially powerful approach, called *homozygosity mapping*, has recently been described for mapping rare recessive diseases, by using the DNA of inbred affected children (24). As long ago as 1902, Garrod (39) noted that a large fraction of his patients with alkaptonuria were the products of marriages between relatives. Almost immediately, Bateson provided the Mendelian explanation that such marriages are excellent ways for a recessive disease allele to become homozygous. Indeed, the rarer the disease, the greater will be the proportion of consanguineous marriages among the parents of affecteds: if disease alleles have population frequency q, the chance of homozygosis is proportional to q^2 in the general population, but proportional to q in an inbred. a substantial chromosomal region surrounding it becomes homozygous by descent as well. In the case of the child of a first-cousin marriage, this region of homozygosity by descent has median length of about 28 cM. This suggests a way to find the disease locus: search for a region that is consistently homozygous by descent in inbred affected children.

If it were possible to recognize homozygosity by descent with perfect accuracy, one could detect linkage to a disease with only three inbred affected progeny of first-cousin marriages. For, the chance that the correct region will be homozygous by descent in such a child is 1 (ignoring the rare circumstance that a second allele is segregating) and the chance that an incorrect region is homozygous by descent is 1/16, the coefficient of inbreeding for a first-cousin marriage. Thus, the odds ratio in favour of such a region being correct is 16:1. If the region is found to be homozygous by descent in three independent such inbred children, the odds ratio is then 4096:1, exceeding the conventional threshold of 1000:1.

Of course, it is not possible to recognize homozygosity by descent with perfect certainty. However, one can use a dense RFLP linkage map to search for a region in which many adjacent RFLPs have become homozygous. The power of the method will hinge on the density of RFLPs used and their degree of polymorphism (i.e. the chance that they would be found homozygous in the general population). *Figure 6* shows the number of inbred first- or second-cousin progeny required for mapping a homogeneous recessive disease. Using an RFLP map with moderately polymorphic loci every 5 cM, fewer than 20 inbred children should suffice for detecting linkage. For relatively rare recessive disorders, homozygosity mapping may in fact be the method of choice for linkage mapping.

Although such a detailed RFLP linkage map is not yet available, the current pace of progress makes it likely that it will exist within a few years at most. Collecting DNAs from affected first- and second-cousin progeny thus already seems advisable. (More distant degrees of inbreeding turn out to be less useful, because an even denser map is needed to detect the smaller region of homozygosity by descent and because there is an increased chance that the cause of the disease might not be due to homozygosity by descent but the introduction into the pedigree of a second disease allele.) Geneticists interested in linkage mapping may wish to pay special attention to those countries in which close-relative marriages are still common. To cite one example, one-third of all Hindu marriages in Andhra Pradesh state in India are between an uncle and niece (24).

10. CONCLUSION

The proposal (1) that RFLPs could serve as an abundant supply of polymorphic genetic markers has clearly revolutionized the study of human genetics, making it possible to map and eventually to clone genes underlying human traits and diseases in much the same manner as in experimental organisms—provided that they show simple modes of inheritance and are relatively common. However, human traits which have more complex modes of inheritance or which are relatively rare require special attention since one cannot use the traditional experimental approaches, involving arranging crosses at will.

By exploiting the full power of a complete RFLP linkage map of the human genome, one can devise alternative approaches to overcoming such genetic complications. Given the current rapid progress in RFLP mapping, the existence of such a map in the next year or two seems assured. In planning future linkage studies, investigators may thus begin to assume the availability of an RFLP linkage map in considering what steps they will take to overcome genetic complexities. Several approaches are described above.

As the reader will have noted, the 'optimal' design of a linkage study will depend to some extent on the precise nature of the genetic complexities to be surmounted (e.g. the number of different loci causing a heterogeneous disorder; the rate of phenocopies). Often, these details cannot be known with precision in advance. One ought to begin a linkage study by enumerating the likely range of complexities, based on the available evidence. One should then determine the available population resources (including large pedigrees, isolated populations, inbred children) and medical resources (including clinical methods for splitting phenotypes) which could be used to simplify the situation. One must then compute the expected number of families of various types needed to map the trait, based on the assumptions about complexities. Finally, one collects the families, prepares DNA and applies RFLP probes at a large number of loci. If the assumptions about the genetic aetiology of the disease are correct, one will stand a good chance of detecting linkage. If the trait is in fact more complicated than assumed, linkage will not be detected. A negative result to a complete genome search will at least prove that the disease is more complex than had been assumed.

By exploiting the full power of an RFLP linkage map of the human genome, it is thus possible to extend the study of human inheritance beyond traits showing simple Mendelian inheritance. A wide variety of medical questions, as well as the general study of mammalian biology, should prove amenable to such explorations.

11. ACKNOWLEDGEMENTS

The author warmly acknowledges his debt to David Botstein in collaboration with whom all of the methods described above were developed. I am grateful for the support of the National Science Foundation (DCB-8611317), the System Development Foundation (G612) and the Whitehead Institute for Biomedical Research.

12. REFERENCES

1. Botstein,D., White,R.L., Skolnick,M. and Davis,R.W. (1980) *Am. J. Hum. Genet.*, **32**, 314.
2. Willard,H., Skolnick,M., Pearson,P.L. and Mandel,J.L. (1985) *Cytogenet. Cell. Genet.*, **40**, 360.
3. Nakamura,Y., Leppert,M., O'Connell,P., Wolff,R., Holm,T., Culver,M., Martin,C., Fujimoto,E., Hoff,M., Kumlin,E. and White,R. (1987) *Science*, **235**, 1616.
4. Conference Proceedings, Human Gene Mapping 9 (1988) *Cytogenet. Cell Genet.*, **46**, 1.
5. Davis,K.E., Pearson,P.L., Harper,P.S., Murray,J.M., O'Brien,T., Sarfarazi,M. and Williamson,R. (1983) *Nucleic Acids Res.*, **11**, 2303.

6. Gusella,J.F., Wexler,N.S., Conneally,P.M., Naylor,S.L. Anderson,M.A., Tanzi,R.E., Watkins,P.C., Ottina,K., Wallace,M.R., Sakaguchi,A.Y., Young,A.B., Shoulson,I., Bonilla,E. and Martin,J.B. (1983) *Nature*, **306**, 234.

7. Tsui,L.-C., Buchwald,M., Barker,D., Braman,J.C., Knowlton,R., Schumm,J.W., Eiberg,H., Mohr,J., Kennedy,D., Plavsic,N., Zsiga,M., Markiewicz,D., Akots,G., Brown,V., Helms,C., Gravius,T., Parker,C., Rediker,K. and Donis-Keller,H. (1985) *Science*, **230**, 1054.

8. Knowlton,R.G., Cohen-Haguenauer,O., Van Cong,N., Frezal,J., Brown,V.A., Barker,D., Braman,J.C., Schumm,J.W., Tsui,L.-C., Buchwald,M. and Donis-Keller,H. (1985) *Nature*, **318**, 381.

9. Wainwright,B.J., Scambler,P.J., Schmidtke,J., Watson,E.A., Law,H.Y., Farrall,M., Cooke,H.J., Eiberg,H. and Williamson,R. (1985) *Nature*, **318**, 384.

10. White,R., Woodward,S., Leppert,M. O'Connelll,P., Hoff,M., Herbst,J., Lalouel,J.-M., Dean,M. and Vande Woude,G. (1985) *Nature*, **318**, 382.

11. Reeders,S.T., Breuning,M.H., Davies,K.E., Nicholls,R.D., Jarman,A.P., Higgs,D.R., Pearson,P.L. and Weatherall,D.J. (1985) *Nature*, **317**, 542.

12. Cavenee,W.K., Hansen,M.F., Nordernskjoid,M., Kock,E., Maumenu,I., Squire,J.A., Phillips,R.A. and Gallie,B.L. (1985) *Science*, **228**, 501.

13. St. George-Hyslop,P.H., Tanzi,R.E., Polinsky,R.J., Haines,J.L., Nee,L., Watkins,P.C., Myers,R.H., Feldman,R.G., Pollen,D., Drachman,D., Growdon,J., Bruni,A., Foncin,J.-F., Salmon,D., Frommelt,P., Amaducci,L., Sorbi,S., Piacentini,S., Stewart,G.D., Hobbs,W.J., Conneally,P.M. and Gusella,J.F. (1987) *Science*, **235**, 885.

14. Egeland,J.A., Gerhard,D.S., Pauls,D.L., Sussex,J.N., Kidd,K.K., Allen,C.R., Hostetter,A.M. and Housman,D.E. (1987) *Nature*, **325**, 783.

15. Barker,D., Wright,E., Nguyen,K., Cannon,L., Fain,P., Goldgar,D., Bishop,D.T., Carey,J., Baty,B., Kivlin,J., Willard,H., Waye,J.S., Greig,G., Leinwand,L., Nakamura,Y., O'Connell,P., Leppert,M., Lalouel,J.-M., White,R. and Skolnick,M. (1987) *Science*, **236**, 1100.

16. Seizinger,B.R., Rouleau,G.A., Ozelius,L.J., Lane,A.H., Faryniarz,A.G., Chao,M.V., Huson,S., Korf,B.R., Parry,D.M., Pericak-Vance,M.A., Collins,F.S., Hobbs,W.J., Falcone,B.G., Iannazzi,J.A., Roy,J.C., St George-Hyslop,P.H., Tanzi,R.E., Bothwell,M.A., Upadhyaya,M., Harper,P., Goldstein,A.E., Hoover,D.L., Bader,J.L., Spence,M.A., Mulvihill,J.J., Aylsworth,A.S., Vance,J.M., Rossenwasser,G.O.D., Gaskell,P.C., Roses,A.D., Martuza,R.L., Breakfield,X.O. and Gusella,J.F. (1987) *Cell*, **49**, 589.

17. Mathew,C.G.P., Chin,K.S., Easton,D.F., Thorpe,K., Carter,C., Liou,G.I., Fong,S.-L.,Bridges,C.D.B., Haak,H., Kruseman,A.C.N., Schifter,S., Hansen,H.H., Telenius,H., Telenius-Berg,M. and Ponder,B.A.J. (1987) *Nature*, **328**, 527.

18. Simpson,N.E., Kidd,K.K., Goodfellow,P.J., McDermid,H., Myers,S., Kidd,J.R., Jackson,C.E., Duncan,A.M.V., Farrer,L.A., Brasch,K., Castiglione,C., Genel,M., Gertner,J., Greenberg,C.R., Gusella,J.F., Holden,J.J.A. and White,B.N. (1987) *Nature*, **328**, 528.

19. Bodmer,W.F., Bailey,C.J., Bodmer,J., Bussey,H.J.R., Ellis,A., Gorman,P., Lucibello,F.C., Murday,V.A., Rider,S.H., Scambler,P., Sheer,D., Solomon,E. and Spurr,N.K. (1987) *Nature*, **328**, 614.

20. Donis-Keller,H., Green,P., Helms,C., Cartinhour,S., Weiffenbach,B., Stephens,K., Keith,T.P., Bowden,D.W., Smith,D.R., Lander,E.S., Botstein,D., Akots,G., Rediker,K.S., Gravius,T., Brown,V.A., Rising,M.B., Parker,C., Powers,J.A., Watt,D.E., Kauffman,E.R., Bricker,A., Phipps,P., Muller-Kahle,H., Fulton,T.R., Ng,S., Schumm,J.W., Braman,J.C., Knowlton,R.G., Barker,D.F., Crooks,S.M., Lincoln,S., Daly,M. and Abrahamson,J. (1987) *Cell*, **51**, 3197.

21. White,R., Lalouel,J.M., O'Connell,P., Nakamura,Y., Leppert,M. and Lathrop,M. (1987) Booklet printed by Howard Hughes Medical Institute.

22. Lander,E.S. and Botstein,D. (1986) *Cold Spring Harbor Symp. Quant. Biol.*, **51**, 49.

23. Lander,E.S. and Botstein,D. (1986) *Proc. Natl. Acad. Sci. USA*, **83**, 7353.

24. Lander,E.S. and Botstein,D. (1987) *Science*, **236**, 1567.

25. Ott,J. (1986) In *Human Genetic Diseases—A Practical Approach*. Davies,K.E. (ed.), IRL Press, Oxford, p. 19.

26. Morton,N.E. (1955) *Am. J. Hum. Genet.*, **7**, 277.

27. Blass,J.P. and Gibson,G.E. (1977) *N. Engl. J. Med.*, **297**, 1367.

28. Stanbury,J.B., Wyngaarden,J.B., Frederickson,D.S., Goldstein,J.L. and Brown,M.S. (1983) *The Metabolic Basis of Inherited Disease*. McGraw-Hill, New York.

29. Morton,N. (1956) *Am. J. Hum. Genet.*, **8**, 80.

30. Bird,T.D., Ott,J., Giblett,E.R, Chnace,P.F., Sumi,S.M. and Kraft,G.H. (1983) *Ann. Neurol.*, **14**, 679.

31. Dyck,P.J., Ott,J., Moore,S.B., Swanson,C.J. and Lambert,E.H. (1983) *Mayo Clin. Proc.*, **58**, 430.

32. Hodgkinson,S., Sherrington,R., Gurling,H., Marchbanks,R., Reeders,S., Mallet,J., McInnis,M., Petursson,H. and Brynjolfsson,J. (1987) *Nature*, **325**, 805.

33. Detera-Wadleigh,S.D., Berrettini,W.H., Goldin,L.R., Boorman,D., Anderson,S. and Gershon,E. (1987)

Nature, **325**, 806.

34. Jaspers,N.G.J., DeWit,J., Regulski,M.R. and Bootsma,D. (1982) *Cancer Res.*, **42**, 335.
35. Jaspers,N.G.J., Painter,R.B., Paterson,M.C., Kidson,C. and Inoue,T. (1985) In *Ataxia-telangiectasia: Genetics, Neuropathology, and Immunology of a Degenerative Disease of Childhood*. Gatti,R.A. and Swift,M. (eds), A.R.Liss, New York, p. 147.
36. Keijzer,W., Jaspers,N.G.J., Abrahams,P.J., Taylor,A.M.R., Arlett,C.F., Takebe,H., Kimmont,P.D.S. and Bootsma,D. (1979) *Mutat. Res.*, **62**, 183.
37. Trevor-Roper,P.D. (1952) *Br. J. Opthalmol.*, **36**, 107.
38. Rouleau,G.A., Wertelecki,W., Haines,J.L., Hobbs,W.J., Trofatter,J.A., Seizinger,B.R., Martuza,R.L., Superneau,D.W., Conneally,P.M. and Gusella,J.F. (1987) *Nature*, **329**, 246.
39. Garrod,A.E. (1902) *Lancet*, **11**, 1616.

INDEX